OUR UNIVERSE-- GOD'S INCEPTION, MAN'S PERCEPTION

Understanding Reality

Dr. Matthias Magoola

Chicago University, Inc.

Copyright © 2024 Matthias Magoola

All rights reserved

The characters and events portrayed in this book are fictitious. Any similarity to real persons, living or dead, is coincidental and not intended by the author.

No part of this book may be reproduced, or stored in a retrieval system, or transmitted in any form or by any means, electronic, mechanical, photocopying, recording, or otherwise, without express written permission of the publisher.

ISBN-13: 9798336218350
ISBN-10: 833621835

Cover design by: Art Painter
Library of Congress Control Number: 2018675309
Printed in the United States of America

His Excellency Yoweri Kaguta Museveni Tibuhaburwa, President of the Republic of Uganda, for his zeal to make Africa a continent of innovations.

I think therefore, I am.

                                                RENE DESCARTES

CONTENTS

Title Page
Copyright
Dedication
Epigraph
Preface
About the Book

Chapter 1: What is the Universe?	1
Chapter 2: Do We Exist, or Is It an Imagination?	19
Chapter 3: What Was There Before the Big Bang, or Was There a Big Bang?	41
Chapter 4: Einstein Understood	58
Chapter 5: Quantum Mechanics for Laymen	78
Chapter 6: From Energy to Living Entity	91
Chapter 7: Quantum Entanglement Within the Brain	99
Chapter 8: Do We Need to Age?	113
Chapter 9: What Would We Talk About in the Year 3024?	125
Chapter 10: Epilogue: Reconciling Faith and	152

science

References: Chapter 1	160
references: Chapter 2	162
references: Chapter 3	164
references: Chapter 4	166
references: Chapter 5	167
referebces: Chapter 6	170
references: Chapter 7	173
references: Chapter 8	176
references: Chapter 9	178
references: Chapter 10	180

PREFACE

In every era, humans grapple with profound questions: Is the story of creation a myth? Is the concept of a divine God or supreme being a hoax? Is divine worship necessary? Does death signify the end of existence? Are holy scriptures truth or fiction? Does science contradict or disprove these sacred texts? Is human life merely a fleeting moment between an eternal darkness that preceded birth and follows death? These are questions that any intelligent person is bound to ask once they reach serious contemplation.

In my search for answers, I meticulously studied the entire Bible multiple times, read the Holy Quran, and delved into numerous spiritual and scientific texts on evolution, intending to discover who we are. Where did we come from? How did the universe and everything within it come into existence? How did the universe start? After reading the Bible, I initially felt content and believed all my questions had been answered.

However, as I explored scientific texts on evolution, what we consider "reality" began challenging my faith. As a devout Christian, scientist, and researcher, I encountered findings that threatened my beliefs: dinosaur fossils dating back over seventy million years, rocks billions of years old, human skeletons over two million years old, and ancestors' fossils dating back more than five million years. The age of seawater has been confirmed to be billions of years old. These facts can be observed in museums around the world, such as the one in Washington, D.C., and they stand in stark contrast to the biblical account of creation, which suggests the universe was created in six days, less than 10,000 years ago, based on genealogies provided in the Bible.

These contradictions have led many to abandon their faith, believing that trusting in God or the story of creation is futile since science has seemingly proven it wrong. I sought explanations from priests, theologians, and philosophers but needed more satisfactory answers. Yet, we must address these questions to maintain faith in the modern world. Some argue that the Bible, Quran, and other spiritual texts are not scientific works, but the Bible declares that all scripture is eternally true. So, when scientific discoveries contradict scripture, it becomes a stumbling block for many, leading them to abandon faith or embrace atheism.

Faith and reason serve different roles in our lives. Reason deals with the known and the routine, while faith explores the unknown and ventures beyond limitations. While reason grounds us, faith provides comfort and allows us to transcend the laws of nature. Faith that arises out of fear or greed is fragile, whereas faith born from love, like that between a mother and child or a master and disciple, is unbreakable. Faith is a source of strength, stability, and happiness, bringing calmness and love. True faith is not about doing a favor to God but about gaining instant strength and peace from within.

The Bible's Genesis, part of the Torah, was likely written around the 5th century B.C. Mainstream biblical scholars view Genesis as largely mythological rather than historical, reflecting humanity's understanding at a time when survival preoccupied their minds. It is unlikely that people then could have grasped scientific concepts like the Earth's rotation around the Sun, much as Galileo's observations were met with disbelief and persecution centuries later. Holy texts aim to increase our faith and understanding of our relationship with the Creator, not to provide scientific explanations.

Humanity often overestimates its importance and knowledge, dismissing what we cannot see or understand as nonexistent or a hoax. Yet, many things we rely on daily, like oxygen, bacteria, cells, and even subatomic particles, are invisible to the naked eye. Despite our technological advances, we have yet to unravel many mysteries of the universe, such as quantum entanglement, the nature of time, or the forces that bind particles together. Even with powerful tools, many phenomena remain beyond our comprehension, suggesting the presence of a higher power.

The Bible states in John 21:25, "Now, there are many other things that Jesus did. If they were all written down one by one, I suppose the whole world could not hold the books that would be written." Similarly, 2 Peter 3:8 and Psalms 90:4 emphasize the divine perspective on time, suggesting that a day to God is like a thousand years to humans, echoing Einstein's notion of time as a fluid concept. The Quran also echoes this sentiment, stating that if all the trees were used to describe the Creator, they would be exhausted before the task was complete.

Throughout history, the divine presence has been manifesting in ways that defy human understanding, as exemplified by Eucharistic miracles or the lives of saints whose mediations have led to medically inexplicable events. Such occurrences, which science cannot fully explain, remind us of our limitations and the presence of a higher power. They challenge us to remain humble and acknowledge that our knowledge is finite and our existence fragile.

In conclusion, I urge humanity to remain humble and seek God while He can still be found. Our plans, knowledge, and achievements are limited without the divine guidance of God. Whether we live or die, we remain within His domain, and our only choice is to obey and do His will to live eternally or reject Him and face eternal consequences. The lessons of quantum mechanics and scientific discoveries should not shake our faith but reinforce it, highlighting the vastness of God's creation and our place within it. With all its complexities and mysteries, the universe only deepens our appreciation for the Creator's infinite power and wisdom.

Thank you for reading. May God bless you. Gloria in Excelsis Deo.

Matthias Magoola, Kampala, Uganda

ABOUT THE BOOK

The book focuses on bridging the often-contentious divide between modern science and religious belief, offering readers a comprehensive understanding of how the universe and life itself can be explained through scientific principles while acknowledging the deeply ingrained human inclination toward spiritual or religious interpretations.

Chapter 1 introduces the reader to the fundamental concepts of cosmology, exploring the universe's origins from the Big Bang to the present day. It lays the groundwork for understanding the complex processes that have shaped our cosmos, emphasizing the importance of scientific inquiry in unraveling the mysteries of existence.

Chapter 2 delves into the formation and evolution of galaxies, stars, and planetary systems, focusing on our solar system. This chapter highlights the marvels of astrophysics and how, through science, we have come to understand the dynamic and interconnected nature of the universe.

Chapter 3 shifts the focus to the origins of life, exploring the latest scientific theories about how life began on Earth. The chapter discusses the intricate processes of chemical evolution and the emergence of biological complexity, demonstrating how life, as we know it, is a product of natural forces.

Chapter 4 tackles the development of consciousness and the human brain, examining how our cognitive abilities have evolved over millions of years. It addresses the role of neuroscience in explaining human behavior and thought processes, challenging traditional views that attribute consciousness to a divine source.

Chapter 5 critically analyzes the relationship between science and religion, tracing the historical conflicts and ongoing debates that have shaped this complex interaction. It argues that while science seeks to explain the natural world through empirical evidence, religion often provides meaning and purpose that science alone cannot offer.

Chapter 6 explores the ethical implications of scientific advancements, particularly in genetics, artificial intelligence, and biotechnology. It discusses these technologies' moral dilemmas and how they challenge traditional religious and philosophical beliefs.

Chapter 7 examines the role of science in addressing global challenges such as climate change, disease, and resource depletion. It emphasizes the necessity of scientific literacy in navigating these issues and the potential conflicts that arise when religious ideologies oppose scientific solutions.

Chapter 8 reflects on the limits of science, acknowledging that while science has made remarkable strides in understanding the universe, there remain questions that may never be fully answered. It discusses the philosophical implications of these limits and the space they leave for religious or spiritual interpretations.

Chapter 9 looks to the future of science and religion, considering how both might evolve in response to discoveries and societal changes. It suggests that while the conflict between science and religion will likely persist, there may be opportunities for dialogue and mutual understanding.

Finally, Chapter 10 offers a synthesis of the book's themes, urging readers to embrace the complexities of modern science even when they challenge deeply held beliefs. It argues that understanding science is essential for navigating the contemporary world. Still, it also recognizes that human nature may never fully reconcile the tension between scientific rationality and spiritual faith.

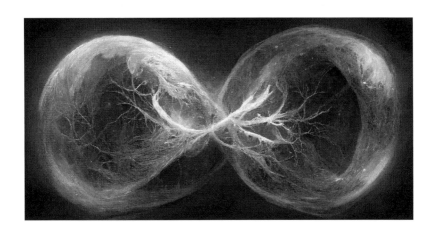

CHAPTER 1: WHAT IS THE UNIVERSE?

The universe encompasses everything—space, time, matter, and energy. It is the vast, almost incomprehensible expanse where all known physical laws operate, stretching across unimaginable distances and operating on timescales that defy human comprehension. Humanity has been captivated by the mysteries of the universe for millennia, from ancient civilizations that looked up at the stars and spun myths and stories to modern-day scientists who use advanced telescopes and particle accelerators to unravel its secrets.

The quest to understand the universe is fundamental to human curiosity, driving us to explore, question, and seek meaning.

Understanding the universe requires grappling with concepts that stretch the limits of our imagination. The scales involved are both incomprehensibly large and infinitesimally small. The universe operates on timescales that span billions of years and at speeds that challenge our understanding of physics. Yet, it also contains the everyday world we experience, governed by the same fundamental laws that apply to the farthest reaches of space. In this chapter, I will explore the universe's structure, scale, and evolution, delving into the fundamental forces that govern it and examining humanity's place within this vast expanse. Our journey will take us from the formation of the first stars and galaxies to the potential fate of the universe billions of years from now. I will also consider the philosophical implications of our understanding of the cosmos and what it means for our place in the universe.

Structure of the Universe

The universe's structure is a complex and dynamic system composed of various levels of organization. From the smallest particles to the most significant cosmic structures, each component plays a crucial role in the overall makeup of the universe.

Cosmic Structures

Stars are the fundamental building blocks of the universe, massive, luminous spheres of plasma that generate energy through nuclear fusion in their cores. This process fuses hydrogen into helium, releasing tremendous energy through light and heat. Stars vary greatly in size, temperature, and lifespan. The smallest stars, such as red dwarfs, can burn for trillions of years, while the largest stars, known as supergiants, may last only a few million years before exploding in a supernova. Stars form from the gravitational collapse of gas clouds in space. As the gas collapses, it heats up and begins the process of nuclear fusion. The life cycle of a star depends on its mass. Like our Sun, low-mass stars will eventually exhaust their nuclear fuel and shed their outer layers, leaving behind a dense core known as a white dwarf. High-mass stars will end their lives in violent supernovae, which can result in the formation of neutron stars or black holes.

Galaxies are vast systems containing billions or even trillions of stars, along with gas, dust, and dark matter, all bound together by gravity. Galaxies come in various shapes, including spiral, elliptical, and irregular forms. The Milky Way, our home galaxy, is a barred spiral galaxy with a supermassive black hole at its center. Galaxies are not isolated; they interact with each other through gravitational forces. These interactions can lead to mergers, where two galaxies combine to form a giant galaxy, or tidal interactions, where the gravitational pull of one galaxy distorts another. These processes have played a crucial role in the evolution of galaxies over billions of years.

Galaxies group into clusters bound by gravity, sometimes containing hundreds or thousands of galaxies. They are some of the most significant structures in the universe. The Milky Way is part of the Local Group, a small cluster of galaxies that includes Andromeda and the Triangulum Galaxy. Superclusters are even larger structures that contain multiple galaxy clusters. They are connected by vast filaments of dark matter and separated by enormous voids where very little matter exists. The cosmic web I will explore later is the universe's large-scale structure formed by these interconnected superclusters and voids.

Large-Scale Structure

The universe's large-scale structure resembles a cosmic web, with galaxies and clusters forming filaments and voids. This structure emerged from the gravitational collapse of matter over billions of years, directly resulting from the initial density fluctuations in the early universe, which were amplified by gravitational forces. The cosmic web comprises filaments, dense regions of galaxies and dark matter, and voids, vast empty spaces between the filaments. The filaments are connected at nodes, where multiple filaments intersect, forming clusters and superclusters of galaxies. The cosmic web is an intricate and interconnected system that reflects the large-scale distribution of matter in the universe.

Dark matter is an invisible substance that does not emit, absorb, or reflect light, making it undetectable by traditional telescopes. However, its presence is inferred from its gravitational effects on visible matter, radiation, and the universe's large-scale structure. Dark matter is thought to account for about 27% of the universe's mass-energy content and is crucial in shaping the cosmic web. It is believed to be composed of exotic particles that interact weakly with ordinary matter, forming the "gravitational glue" that holds galaxies and clusters together, preventing them from flying apart due to their rotational speeds. The search for dark matter particles is one of modern physics's most important research areas.

Even more mysterious than dark matter is dark energy, which makes up about 68% of the universe's mass-energy content. Dark energy is responsible for the universe's accelerated expansion, as observed in distant supernovae and the large-scale structure of the cosmos. The nature of dark energy remains one of the greatest mysteries in cosmology. It is often associated with the cosmological constant, a term Einstein introduced in his general relativity equations. However, its true nature is still unknown, and understanding dark energy is one of the significant challenges in modern cosmology. It could be a property of space, a new form of energy, or something entirely different.

The Scale of the Universe

The scale of the universe is vast beyond human comprehension, encompassing the smallest subatomic particles and the most significant cosmic structures, spanning distances and timescales that challenge our understanding of reality.

Astronomical distances are often measured in light-years and parsecs. A light-year is the distance that light travels in one year, about 5.88 trillion miles (9.46 trillion kilometers). For example, the nearest star to Earth, Proxima Centauri, is about 4.24 light-years away. A parsec is another unit of distance used in astronomy, equal to about 3.26 light-years. It is based on the parallax angle, which is the apparent shift in position of a nearby star against the background of distant stars as observed from Earth at opposite points in its orbit. Measuring distances in the universe is a challenging task. Astronomers use various methods, including parallax, standard candles (such as Cepheid variables and Type Ia supernovae), and redshift, to estimate the distances to stars, galaxies, and other cosmic objects. These measurements are crucial for understanding the scale and structure of the universe.

Within our solar system, distances are commonly measured in astronomical units (AU). One AU is the average distance between Earth and the Sun, about 93 million miles (150 million kilometers). This unit is convenient for measuring distances between objects in the solar system, such as the distance from the Sun to Mars (1.5 AU) or the Sun to Pluto (about 40 AU). The use of AU helps to simplify the comparison of distances within the solar system, making it easier to understand the relative positions of planets and other celestial bodies. For example, the distance from the Sun to the Oort Cloud, a distant region of icy objects, is estimated to be about 50,000 AU.

The universe spans a vast range of sizes, from the smallest subatomic particles to the observable universe, about 93 billion light-years in diameter. Subatomic particles, such as quarks and electrons, are the building blocks of matter and operate at scales smaller than a trillionth of a meter. At the other end of the scale, the observable universe encompasses all the galaxies, stars, and other cosmic structures we can see or detect from Earth. The vastness of the observable universe challenges our understanding of space and time, as it operates on scales that are difficult to comprehend. The observable universe is only a small portion of the entire universe, which could be much larger or even infinite. The true extent of the universe remains speculative, as regions beyond the observable universe are beyond our current observational capabilities.

The observable universe is the portion of the universe that we can see, limited by the speed of light and the universe's age. The cosmic horizon represents the maximum distance from which light has had time to reach us since the beginning of the universe, about 13.8 billion years ago. Beyond this horizon, light from distant objects has not yet reached us, rendering those regions invisible. The cosmic horizon constantly expands as the universe ages, allowing us to see more distant objects over time. However, the universe's expansion also means that some regions are moving away from us faster than the speed of light, effectively placing them beyond our reach forever. The concept of the cosmic horizon has essential implications for our understanding of the universe's structure and evolution. It defines the limits of our observable universe and highlights the potential existence of regions that we may never be able to observe or study.

The Life Cycle of the Universe

The universe has a dynamic life cycle, from its birth in the Big Bang to its potential future fates. Understanding this life cycle is key to comprehending the evolution of the cosmos and the processes that have shaped its structure over billions of years.

The prevailing theory of the universe's origin is the Big Bang theory, which posits that the universe began as an extremely hot and dense point about 13.8 billion years ago. This singularity rapidly expanded, cooling and forming the basic building blocks of matter. The Big Bang marks the beginning of space and time and the birth of the universe as we know it. The Big Bang theory is supported by several lines of evidence, including the cosmic microwave background radiation, the observed expansion of the universe, and the abundance of light elements such as hydrogen and helium. These observations support the idea that the universe began in a hot, dense state and has expanded ever since. The initial conditions of the Big Bang were incredibly extreme, with temperatures reaching trillions of degrees and densities far exceeding anything we see today. As the universe expanded, it cooled rapidly, allowing particles to form and interact, eventually forming atoms, stars, and galaxies.

In the first few seconds after the Big Bang, fundamental particles like quarks, electrons, and neutrinos formed. These particles eventually combine to form protons and neutrons, which are the building blocks of atomic nuclei. The particle formation process in the early universe was governed by the fundamental forces of nature, including the strong and weak nuclear forces, electromagnetism, and gravity. These forces determined how particles interacted and combined, leading to the formation of the first atoms. As the universe expanded and cooled, these particles underwent further interactions, forming more complex structures, such as atomic nuclei and neutral atoms. This period of the universe's history is known as nucleosynthesis, and it played a crucial role in determining the composition of the universe we see today.

Nucleosynthesis is the process by which light elements such as hydrogen, helium, and trace amounts of lithium were synthesized in the early universe. This process occurred within the first few minutes after the Big Bang, as the universe cooled and the conditions became suitable for nuclear reactions. The abundance of light elements observed in the universe today is consistent with the predictions of Big Bang nucleosynthesis. The relative proportions of hydrogen, helium, and lithium provide important clues about the conditions in the early universe and support the idea that these elements were formed during the first few minutes of the universe's existence. Nucleosynthesis did not produce heavier elements like carbon, oxygen, or iron. These elements were formed later in the cores of stars through nuclear fusion and in supernovae, where the intense temperatures and pressures created more complex elements.

As the universe expanded and cooled further, electrons combined with protons and neutrons to form neutral atoms, allowing light to travel freely and creating the cosmic microwave background radiation. This period, known as recombination, occurred about 380,000 years after the Big Bang and marked a significant milestone in the universe's evolution. The formation of neutral atoms allowed the universe to become transparent to radiation, as photons could travel through space without being constantly scattered by free electrons. This transition from an opaque to a transparent universe created the cosmic microwave background radiation, which we can still observe today. The cosmic microwave background radiation provides a snapshot of the universe at this early stage, revealing the distribution of matter and energy and providing vital clues about the universe's structure and evolution.

Over millions of years, gravity pulled matter together, forming the first stars and galaxies. These stars created heavier elements through nuclear fusion and supernovae, seeding the universe with the building blocks for planets and life. The formation of the first stars, known as Population III stars, marked the beginning of the era of galaxy formation. These massive and short-lived stars ended their lives in powerful supernovae that enriched the surrounding gas with heavy elements. The star formation process continues today, with new stars born in regions of dense gas and dust known as molecular clouds. The cycle of star formation and death has played a crucial role in shaping the universe's structure and evolution, leading to the formation of galaxies, planets, and life.

If the universe continues to expand indefinitely, it will eventually reach a state where stars burn out and galaxies drift apart, leading to a cold, dark, and dilute universe. This scenario is known as the universe's Big Freeze or heat death. In the Big Freeze scenario, the universe's expansion, driven by dark energy, accelerates. As galaxies move farther apart, the formation of new stars slows down, and existing stars eventually exhaust their nuclear fuel. Over time, black holes and subatomic particles will be the last remnants of a once vibrant universe. The Big Freeze represents the ultimate fate of a universe dominated by dark energy, where entropy increases to the point where no usable energy remains, and all physical processes cease.

If the density of matter in the universe is sufficient to halt its expansion, gravity could eventually cause the universe to collapse back into a singularity, potentially leading to a cyclic model of rebirth. This scenario is known as the Big Crunch. In the Big Crunch scenario, the universe's expansion slows down and eventually reverses, leading to a contraction of space-time. As the universe collapses, galaxies, stars, and planets are drawn closer together, eventually merging into a single, incredibly dense point. The Big Crunch could lead to a new Big Bang, where the universe is reborn in a new cycle of expansion and contraction. This cyclic model of the universe has been proposed as an alternative to a singular beginning and end.

If dark energy's repulsive force increases, it could overcome all other forces, tearing apart galaxies, stars, planets, and eventually atoms. This scenario is known as the Big Rip. In the Big Rip scenario, the universe's expansion accelerates to the point where the fabric of space-time itself is torn apart. As dark energy's influence grows more robust, it disrupts the gravitational forces that bind galaxies, stars, and planets, eventually destroying them. The Big Rip represents a catastrophic end to the universe, where even the smallest particles are torn apart by the relentless expansion of space-time. It is one of the more speculative scenarios for the universe's ultimate fate, but it remains a possibility within the framework of current cosmological theories.

The Laws Governing the Universe

The universe operates according to fundamental laws that govern the behavior of matter and energy. These laws are expressed through the fundamental forces of nature and the physical constants that define the universe's structure and evolution.

Gravity is the force of attraction between masses. It governs the motion of planets, stars, and galaxies and plays a crucial role in forming and evolving cosmic structures. Gravity is described by Einstein's theory of general relativity, which portrays it as the curvature of space-time caused by mass and energy. Gravity is the weakest of the four fundamental forces but has an infinite range and dominates on large scales. It is responsible for forming stars, galaxies, and the universe's large-scale structure. The discovery of gravitational waves, ripples in space-time caused by accelerating masses, has provided new insights into the nature of gravity and its role in the universe.

Electromagnetism is the force between charged particles responsible for electricity, magnetism, and light. It is described by Maxwell's equations, which unify the concepts of electric and magnetic fields. Electromagnetism is fundamental to the structure of atoms and molecules and governs the interactions between charged particles. Electromagnetism has an infinite range and is much stronger than gravity, but opposite charges can cancel it out. It plays a central role in the behavior of matter and energy on both small and large scales, from the interactions of subatomic particles to the behavior of electromagnetic radiation in space.

The strong nuclear force binds protons and neutrons in atomic nuclei, overcoming the electromagnetic repulsion between positively charged protons. It is the strongest of the four fundamental forces but has a concise range, operating only at the scale of atomic nuclei. The strong nuclear force is responsible for the stability of atomic nuclei and the energy released in nuclear reactions, such as fusion in stars and fission in nuclear power plants. It is described by the theory of quantum chromodynamics (QCD), which explains the interactions between quarks and gluons, the fundamental particles that make up protons and neutrons.

The weak nuclear force is responsible for beta decay in radioactive elements, where a neutron transforms into a proton, an electron, and an antineutrino. It plays a role in nuclear reactions within stars and is essential for nucleosynthesis. The weak nuclear force has a concise range and is weaker than the strong nuclear force and electromagnetism. It is described by the electroweak theory, which unifies the weak nuclear force with electromagnetism. The discovery of the Higgs boson in 2012 confirmed the existence of the Higgs field, which gives particles mass and is linked to the weak nuclear force.

Physical constants, such as the speed of light, Planck's constant, and the gravitational constant, are fundamental values that define the universe's behavior. These constants are universal and unchanging, providing the foundation for the laws of physics. The speed of light (c) is the maximum speed at which information can travel in the universe and is central to the theory of relativity. Planck's constant (h) sets the scale for quantum mechanical effects, defining the relationship between energy and frequency. The gravitational constant (G) determines the strength of the gravitational force between two masses. These constants are finely tuned to allow for a stable and life-supporting universe. Small changes in their values could dramatically affect the universe's structure, potentially preventing the formation of stars, galaxies, and life.

The precise values of physical constants have led to the concept of fine-tuning, where the universe appears specifically configured to support life. This observation has prompted debates about the nature of the universe and whether it is the result of chance, necessity, or design. Fine-tuning has also led to exploring the multiverse hypothesis, where our universe is one of many, each with different physical constants. In this view, the fine-tuning we observe may result from our existence in a universe with the right conditions for life.

Physicists strive to develop a Theory of Everything (ToE) that unifies all fundamental forces into a single theoretical framework. Current candidates include string theory and loop quantum gravity, though a complete ToE remains elusive. String theory proposes that the fundamental particles of nature are not point-like objects but tiny, vibrating strings. These strings can vibrate at different frequencies, giving rise to the various particles we observe. String theory also requires the existence of additional spatial dimensions beyond the familiar three. Loop quantum gravity is another approach to unifying gravity with quantum mechanics. It proposes that space-time is composed of discrete loops rather than being continuous. This theory aims to reconcile general relativity with quantum mechanics and could provide insights into the nature of black holes and the early universe. The search for a Theory of Everything is one of the most ambitious goals in modern physics. It represents the quest to understand the fundamental nature of reality and the underlying principles that govern the universe.

Humanity's Place in the Universe

Humanity's place in the universe has fascinated philosophers, scientists, and theologians for centuries. Understanding the cosmos shapes our sense of identity, purpose, and connection to the broader tapestry of existence.

The anthropic principle suggests that the universe's physical constants appear fine-tuned for the emergence of life. This observation raises questions about the nature of the universe and our place within it. The anthropic principle comes in two forms: the weak anthropic principle, which states that the universe must allow for the existence of observers, and the strong anthropic principle, which suggests that the universe is in some way compelled to develop life and consciousness. The fine-tuning of physical constants, such as the gravitational and cosmological constants, has led to debates about whether this results from chance, necessity, or design. Some scientists and philosophers argue that fine-tuning is evidence of a multiverse, where our universe is just one of many, each with different physical laws. Others suggest that fine-tuning may indicate the presence of a higher purpose or design.

The anthropic principle has profound philosophical implications, challenging our understanding of existence and the nature of reality. It raises questions about the role of consciousness in the universe and whether our existence is merely a cosmic accident or part of a larger plan. The anthropic principle also intersects with questions of free will, determinism, and the nature of time. It encourages us to consider our place in the universe as observers and participants in a vast and interconnected cosmic process.

The Fermi paradox questions why, given the vastness of the universe, we have not yet detected signs of extraterrestrial civilizations. Named after physicist Enrico Fermi, the paradox highlights the apparent contradiction between the high probability of extraterrestrial life and the lack of evidence for it. Possible explanations for the Fermi paradox include the rarity of intelligent life, the vast distances between civilizations, and the possibility that advanced civilizations are avoiding contact or are using communication methods we cannot detect. Some theories suggest that civilizations may self-destruct before achieving interstellar communication or exist in forms beyond our current understanding. The Fermi paradox has inspired numerous science fiction stories and serious scientific inquiry. It challenges us to consider the implications of being alone in the universe and the potential for discovering other intelligent beings.

The Drake equation is a probabilistic formula to estimate the number of active, communicative extraterrestrial civilizations in our galaxy. It considers factors such as the rate of star formation, the fraction of stars with planets, the number of planets that could support life, and the likelihood of life developing intelligence and communication. The Drake equation is not meant to provide a precise number but to stimulate discussion and guide research efforts in the search for extraterrestrial life. It highlights the many uncertainties and variables involved in the search and underscores the need for continued exploration and discovery. The search for extraterrestrial life involves various methods, including the study of exoplanets, the search for biosignatures in planetary atmospheres, and the monitoring radio signals for potential messages from other civilizations. The discovery of life beyond Earth would have profound implications for our understanding of the universe and our place within it.

The vastness of the universe can evoke feelings of insignificance, but it can also inspire a sense of wonder and connection to a larger cosmic order. Philosophical and spiritual perspectives vary in interpreting humanity's place in the cosmos. Some philosophers and theologians argue that the vastness of the universe highlights the uniqueness and significance of life on Earth. Others suggest that our small place in the cosmos challenges traditional notions of human importance and encourages a more humble and interconnected view of existence. The study of the universe also intersects with questions of meaning, purpose, and the nature of reality. It invites us to explore our relationship with the cosmos and consider our existence's broader implications. The concept of cosmic evolution, where the universe is seen as a dynamic and evolving system, offers a framework for understanding our place in the cosmos. It suggests that life and consciousness are part of a larger process of cosmic development, with humanity playing a unique role in this unfolding story.

Conclusion

The universe is a vast and dynamic system governed by fundamental forces and laws that shape its structure and evolution. From the smallest subatomic particles to the most significant cosmic structures, the universe operates on scales that challenge our understanding of space and time. The universe's life cycle, from the Big Bang to its potential future fates, provides insights into the processes that have shaped the cosmos over billions of years. Understanding these processes is key to comprehending the universe's past, present, and future. Humanity's place in the universe is a profound philosophical and scientific inquiry topic. The search for extraterrestrial life, the exploration of the cosmos, and the study of the universe's fundamental laws all contribute to our understanding of our role in the cosmic order.

Humanity's quest to comprehend the universe advances scientific knowledge and shapes our sense of identity and place within the cosmos. As we continue to explore and learn, we are reminded of the importance of curiosity, open-mindedness, and the relentless pursuit of knowledge. Continued universe exploration requires interdisciplinary collaboration, drawing from astronomy, physics, mathematics, and philosophy. It also involves the development of new technologies and methods for observing and studying the cosmos.

In all its grandeur and mystery, the universe remains a source of endless fascination and inspiration. Through this journey of discovery, we deepen our understanding of the cosmos and our connection to the broader tapestry of existence. Studying the universe encourages us to reflect on the nature of reality, consciousness's role, and life's significance. It challenges us to think beyond our immediate surroundings and consider our place in a vast and interconnected cosmic order. As we continue to explore the universe, we are reminded of the importance of wonder, curiosity, and the pursuit of knowledge. The universe is not just a physical system but a source of inspiration that invites us to explore the most profound questions of existence.

CHAPTER 2: DO WE EXIST, OR IS IT AN IMAGINATION?

The question of existence has been a central theme in philosophy, science, and religion for millennia. It touches on the essence of reality, the self, and the universe. Are we, as conscious beings, genuinely part of a tangible, objective world, or is our existence merely a product of imagination, perception, or illusion? This chapter delves into the various perspectives and theories regarding the nature of existence, exploring the question's philosophical, scientific, and metaphysical dimensions.

Existence, as we experience it, seems self-evident. We wake up, interact with our environment, and go about our daily lives with the implicit understanding that we and the world around us are real. However, when we question the foundations of this belief, we encounter profound and often unsettling possibilities. Could it be that our perceptions deceive us? Could everything we know and experience be nothing more than a figment of our imagination or a construct of our mind?

Throughout history, philosophers have grappled with these questions, offering a range of theories about the nature of reality and existence. Modern science, too, has contributed to this discourse, with advances in neuroscience, quantum mechanics, and artificial intelligence challenging our traditional views of what it means to exist. This chapter explores these perspectives, seeking to understand the nature of existence and the implications of different theories for our understanding of reality.

Philosophical Perspectives on Existence

Philosophy has long been concerned with the nature of existence, seeking to answer fundamental questions about reality, consciousness, and the self. Several critical philosophical perspectives have shaped our understanding of existence, each offering a unique approach to whether we truly exist or if our existence is an illusion.

Dualism

Dualism, most famously associated with the philosopher René Descartes, posits that reality consists of two distinct substances: the mind (or soul) and the body (or physical world). Descartes' famous declaration,

"Cogito, ergo sum" ("I think, therefore I am"), emphasizes the certainty of the existence of the self as a thinking entity. According to dualism, the mind and body interact but are fundamentally different, with the mind existing as a non-physical substance that can potentially exist independently of the body. Descartes' dualism was a response to the skepticism of his time, which questioned the reliability of the senses and the existence of the external world. By asserting the existence of the self as a thinking being, Descartes sought to establish a foundation for knowledge that could not be doubted. However, dualism raises significant challenges, particularly in explaining how the mind and body interact and how non-physical substances can influence the physical world.

Despite its influence, dualism faces several challenges, particularly from modern neuroscience and materialist philosophies. The idea that the mind and body are separate substances has been difficult to reconcile with evidence from brain science, which shows that mental processes are closely tied to physical brain activity. Additionally, the problem of interaction—how a non-physical mind can influence a physical body—remains a significant obstacle for dualism. Critics of dualism argue that consciousness and mental states are emergent properties of physical processes in the brain rather than being separate, non-physical entities. This perspective, known as physicalism or materialism, contends that everything that exists is ultimately physical and that the brain's workings can fully explain the mind.

Idealism

Idealism is the philosophical view that reality is fundamentally mental or immaterial. According to idealists, what we perceive as the physical world is a construct of the mind or consciousness. As we experience it, the external world is a projection or representation created by the mind. Idealism challenges the assumption that an objective reality is independent of our perceptions, suggesting instead that our consciousness and mental processes shape reality. One of the most famous proponents of idealism was the philosopher George Berkeley, who argued that objects only exist insofar as they are perceived. In Berkeley's view, "to be is to be perceived" (esse est percipi). This means that objects do not exist independently of our perception of them; instead, their existence is contingent on being observed by a conscious mind. Berkeley also believed that God perceives all things, ensuring their continued existence even when no human observes them.

Idealism challenges our conventional understanding of reality, suggesting that the world we experience is not an external, objective reality but a construct of our mind. This has profound implications for our understanding of existence, as it raises questions about the nature of the self, the reliability of our perceptions, and the existence of other minds. Critics of idealism argue that it leads to solipsism, the view that only one's mind is particular to exist. If reality is a mental construct, it becomes difficult to account for the existence of other conscious beings or an objective world. Additionally, idealism faces challenges from empirical science, which relies on the assumption of an objective reality that can be observed and measured independently of individual perception.

Materialism and Physicalism

Materialism, also known as physicalism, is the philosophical view that everything exists is physical. According to materialists, reality consists of matter and energy, and all phenomena, including consciousness and mental states, can be explained in terms of physical processes. Materialism rejects the notion of a non-physical mind or soul, arguing that consciousness is an emergent property of the brain's complex physical activity. Materialism has its roots in ancient philosophy, with figures like Democritus and Epicurus advocating that the universe is composed of indivisible particles (atoms) and that all phenomena, including human consciousness, arise from their interactions. In the modern era, materialism has been bolstered by advances in science, particularly in physics, chemistry, and neuroscience, which have provided increasingly detailed explanations of the physical processes underlying mental and sensory experiences.

While materialism is the dominant view in contemporary philosophy and science, it has challenges. One of the most significant challenges is the "hard problem of consciousness," which refers to the difficulty of explaining how subjective experiences (qualia) arise from physical processes in the brain. Materialism can explain the mechanisms of perception, memory, and cognition, but it struggles to account for the subjective, first-person nature of conscious experience. Additionally, materialism has been criticized for its reductionist approach, which some argue oversimplifies complex phenomena by reducing them to physical interactions. Critics of materialism argue that it fails to capture the richness and complexity of human experience, including the sense of meaning, purpose, and the experience of free will.

Phenomenology

Phenomenology, developed by philosophers such as Edmund Husserl and Martin Heidegger, emphasizes the importance of subjective experience in understanding reality. Phenomenologists argue that reality is not something that exists independently of our experience but is constituted through our lived experience. In this view, consciousness is always consciousness of something, and the act of experiencing is fundamental to the existence of that which is experienced. Phenomenology seeks to explore the structures of consciousness and how we perceive and interact with the world. It emphasizes the importance of intentionality, the idea that consciousness is always directed toward something, whether it be an object, a thought, or an experience. By focusing on the lived experience, phenomenology aims to uncover the underlying structures and meanings that shape our understanding of reality.

Phenomenology challenges the notion of an objective reality that exists independently of our experience. Instead, it suggests that reality is inextricably linked to our perception and interpretation. This has significant implications for our understanding of existence, as it suggests that the world we experience is shaped by our consciousness and that our existence is defined by our experience of the world. Phenomenology also emphasizes the importance of the "lifeworld," the pre-reflective, everyday world of experience that forms the background of all our thoughts and actions. This lifeworld is the foundation of our existence, and phenomenology seeks to uncover its underlying structures and meanings. By focusing on the lived experience, phenomenology offers a unique approach to understanding existence that prioritizes reality's subjective and experiential aspects.

Existentialism

Existentialism is a philosophical movement that emerged in the 19th and 20th centuries, with thinkers like Søren Kierkegaard, Friedrich Nietzsche, Jean-Paul Sartre, and Albert Camus exploring themes of existence, freedom, and the search for meaning. Existentialists argue that existence precedes essence, meaning that individuals are not born with a predetermined purpose or essence but must create their meaning through their choices and actions. Existentialism grapples with the idea that life may be inherently meaningless, and it challenges individuals to confront this possibility and take responsibility for creating their meaning. This involves embracing freedom, making authentic choices, and accepting the consequences of those choices. For existentialists, existence is defined by the individual's experience of the world and their response to life's challenges and uncertainties.

A central theme in existentialism is the concept of the absurd, the idea that there is a fundamental disconnect between the human desire for meaning and the indifferent, meaningless universe. This sense of absurdity can lead to feelings of despair. Still, existentialists argue that it also allows individuals to assert their freedom and create meaning in the face of a meaningless world. Authenticity is another crucial concept in existentialism, referring to living according to one's true self and values rather than conforming to societal expectations or external pressures. For existentialists, authenticity involves acknowledging the absurdity of existence and embracing the freedom and responsibility that come with it. Existentialism challenges individuals to confront the reality of their existence, to make choices that reflect their true selves, and to create meaning in a world that may be indifferent to their struggles. It offers a powerful framework for understanding the human condition and the nature of existence in a world that may be devoid of inherent purpose.

Scientific Perspectives on Existence

While philosophy has long grappled with questions of existence, modern science has also contributed significantly to our understanding of reality and the nature of existence. Advances in neuroscience, physics, and artificial intelligence have challenged traditional views and opened up new possibilities for exploring the fundamental nature of reality.

Neuroscience and the Nature of Consciousness

Neuroscience has provided valuable insights into the relationship between the brain and consciousness, revealing the complex neural processes that underlie perception, cognition, and emotion. Through techniques such as functional magnetic resonance imaging (fMRI), electroencephalography (EEG), and brain stimulation, scientists have been able to map the brain's activity and identify the neural correlates of consciousness. The dominant view in neuroscience is that consciousness arises from the activity of neurons and their connections in the brain. This view, known as the neural correlate theory of consciousness, suggests that specific patterns of neural activity correspond to specific conscious experiences. For example, certain brain areas are associated with visual perception, while others are linked to memory, language, and decision-making.

Neuroscience has also explored the role of the default mode network (DMN), a network of brain regions active at rest and not focused on the external environment. The DMN involves self-referential thinking, daydreaming, and forming our sense of self. Understanding the DMN and other neural networks is crucial for understanding how the brain generates the experience of consciousness and the sense of being a self.

Despite significant advances in understanding the brain, the "hard problem of consciousness" remains a major challenge for science. The hard problem refers to the difficulty of explaining how subjective experiences, or qualia, arise from physical processes in the brain. While neuroscience can explain how the brain processes information and generates behavior, it struggles to account for consciousness's subjective, first-person experience. The hard problem of consciousness raises essential questions about the nature of existence and whether physical processes can fully explain consciousness. Some scientists and philosophers argue that consciousness may require a new, non-reductive framework beyond the current brain understanding. Others suggest that the hard problem may eventually be solved through continued research and the development of new scientific models.

The hard problem also affects our understanding of the self and personal identity. If consciousness is an emergent brain property, what does this mean for our sense of self and continuity over time? Are we the same person from moment to moment, or is the self an ever-changing construct created by the brain? These questions highlight the complexity of understanding existence from a scientific perspective.

Quantum Mechanics and the Nature of Reality

Quantum mechanics, the branch of physics that deals with the behavior of particles at the smallest scales, has revolutionized our understanding of reality. One of the most famous and perplexing aspects of quantum mechanics is wave-particle duality, the idea that particles such as electrons and photons can exhibit both wave-like and particle-like behavior depending on how they are observed. The observer effect, closely related to wave-particle duality, suggests that observation can influence the behavior of quantum particles. In experiments such as the double-slit experiment, particles appear to behave as waves when not observed, creating an interference pattern. However, when the particles are observed, they behave as discrete particles, and the interference pattern disappears. This phenomenon raises questions about the role of the observer in determining the nature of reality and whether reality exists independently of observation.

The implications of wave-particle duality and the observer effect extend beyond quantum mechanics to broader philosophical questions about the nature of existence. If observation influences reality, does this mean our perception shapes the world around us? How do we reconcile the apparent indeterminacy of quantum mechanics with our everyday experience of a stable, predictable world? Quantum entanglement is another intriguing aspect of quantum mechanics that challenges our traditional understanding of existence. When two particles become entangled, their properties become linked so that the state of one particle instantaneously affects the state of the other, regardless of the distance between them. This phenomenon, known as non-locality, seems to violate the principle of locality, which states that objects can only be influenced by their immediate surroundings.

Quantum entanglement has been experimentally confirmed in numerous studies, and it plays a key role in emerging technologies such as quantum computing and quantum cryptography. However, the implications of entanglement for our understanding of reality are profound. Non-locality suggests that the universe may be fundamentally interconnected in ways that defy our classical notions of space and time. Some physicists and philosophers have speculated that quantum entanglement could provide insights into the nature of consciousness and the interconnectedness of all things. If reality is fundamentally non-local, what does this mean for our understanding of existence and the relationship between the mind and the external world?

The many-worlds interpretation (MWI) of quantum mechanics is a radical and controversial theory that suggests that all possible outcomes of a quantum event occur, creating multiple parallel universes. According to MWI, every time a quantum decision is made, the universe splits into various branches, each representing a different outcome. The many-worlds interpretation challenges our traditional understanding of reality by proposing that there are countless parallel universes, each with its version of events. In this view, the reality we experience is just one of many possible realities, and every possible outcome exists in its parallel universe.

The implications of MWI for existence are staggering. If there are infinite parallel universes, what does this mean for our sense of self and identity? Are there countless versions of "us" living different lives in different universes? How do we reconcile the existence of multiple realities with our experience of a single, coherent world? While the many-worlds interpretation remains speculative and controversial, it raises important questions about the nature of existence and the limits of our understanding of reality. It challenges us to think beyond our everyday experience and consider the possibility that reality is far more complex and multifaceted than we can currently comprehend.

Artificial Intelligence and the Simulation Hypothesis

Artificial intelligence (AI) has made significant advances in recent years, with machines now capable of performing tasks that once required human intelligence. From natural language processing and image recognition to autonomous vehicles and game-playing, AI systems have demonstrated remarkable capabilities that challenge our understanding of intelligence and consciousness. The development of AI has raised questions about the nature of consciousness and whether machines can ever possess true consciousness or self-awareness. Some AI researchers argue that consciousness is an emergent property of complex information processing and that sufficiently advanced AI systems could eventually develop consciousness. Others contend that consciousness is inherently tied to biological processes and that machines will never possess true consciousness, no matter how advanced.

The rise of AI also raises ethical and philosophical questions about the nature of existence. If machines can mimic human intelligence and behavior, what does this mean for our understanding of the self and personal identity? How do we distinguish between a conscious being and a highly advanced AI that can simulate consciousness? These questions challenge our traditional views of existence and force us to reconsider what it means to be a conscious being.

The simulation hypothesis is a speculative theory suggesting that our reality is a sophisticated computer simulation created by a highly advanced civilization. According to this hypothesis, what we perceive as the physical world is an artificial construct, and we, along with everything we experience, are part of a simulated environment. The simulation hypothesis has gained attention recently, particularly regarding virtual reality and AI advancements. Proponents of the theory argue that if it is possible for a civilization to create highly realistic simulations, then it is likely that we are living in one, given the vast number of possible simulations compared to the single "real" reality.

The simulation hypothesis raises profound questions about the nature of existence and the reliability of our perceptions. If our reality is a simulation, what does this mean for our sense of self and identity? How do we determine what is "real" in a simulated environment?

The hypothesis also challenges our understanding of consciousness, suggesting that conscious beings could be created and manipulated within a simulated world. While the simulation hypothesis remains speculative and lacks empirical evidence, it has sparked significant debate and discussion within the scientific and philosophical communities. It challenges us to consider the possibility that our reality is not what it seems and that our existence may be far more complex and artificial than we currently understand.

Metaphysical and Religious Perspectives on Existence

In addition to philosophical and scientific perspectives, metaphysical and religious traditions have long grappled with questions of existence, offering their interpretations of reality, consciousness, and the nature of the self. These perspectives provide alternative frameworks for understanding existence beyond the material and empirical world.

The Nature of the Self in Eastern Philosophies

In Hindu philosophy, particularly in the Advaita Vedanta tradition, the self (Atman) is considered to be identical with Brahman, the ultimate reality or universal consciousness. According to Advaita Vedanta, the individual self is not separate from the rest of existence but a manifestation of the same underlying reality. The perception of individuality and separation is considered an illusion (Maya), and spiritual practice aims to realize the true nature of the self as non-dual and one with Brahman. This perspective challenges the conventional understanding of existence as a collection of separate, individual beings. Instead, it suggests that all existence is interconnected and that the true nature of reality is a unified, non-dual consciousness. The realization of this truth is seen as the key to liberation (Moksha) from the cycle of birth, death, and rebirth (Samsara).

Advaita Vedanta has profoundly influenced Eastern philosophy and spirituality, and its teachings continue to be explored and practiced by seekers of truth and self-realization. It offers a perspective on existence that emphasizes unity, interconnectedness, and the transcendence of the ego. In contrast to Hinduism, Buddhism teaches the concept of Anatta (or Anatman), which means "non-self" or "no-self." According to Buddhist philosophy, the self is not a permanent, unchanging entity but a collection of impermanent and interdependent processes (Skandhas). The belief in a permanent self is considered a source of suffering (Dukkha), and Buddhist practice aims to overcome this attachment to the self and realize the truth of Anatta.

The concept of Anatta challenges the conventional understanding of existence as a continuous, self-contained individual. Instead, it suggests that what we perceive as the self is a dynamic and ever-changing process influenced by causes and conditions. The realization of Anatta is central to the Buddhist path of liberation (Nirvana), where one transcends the cycle of suffering and attains enlightenment. Buddhism's emphasis on impermanence, interdependence, and non-self offers a unique perspective on existence that encourages a deeper understanding of the nature of reality and the self. It challenges us to let go of our attachment to the ego and to embrace a more fluid and interconnected view of existence.

Existence in Abrahamic Religions

In the Abrahamic religions (Judaism, Christianity, and Islam), existence is often understood in terms of divine creation and purpose. According to these traditions, the universe and all living beings were created by a transcendent and personal God with a purpose and plan for creation. Human beings are seen as created in the image of God, endowed with free will and the capacity for moral and spiritual growth. The concept of existence in Abrahamic religions is closely tied to the belief in a personal relationship with God and the idea of life as a journey toward fulfilling a divine purpose. Existence is not seen as an accident or a mere product of physical processes but as a meaningful and purposeful reality shaped by divine intention. This perspective provides a sense of identity, purpose, and hope, as believers are encouraged to live following God's will and to seek a deeper connection with the divine.

The belief in an afterlife is also central to the Abrahamic understanding of existence. In these traditions, existence continues beyond physical death, with the soul either entering into eternal communion with God or facing judgment and consequences based on one's actions in life. This belief in an afterlife provides a moral framework for existence and emphasizes the importance of living a righteous and faithful life. In addition to the mainstream teachings of the Abrahamic religions, mystical traditions within these faiths emphasize the possibility of direct, personal experience of God or the divine. Mystics seek to transcend the ordinary, material world and attain a union or communion with the divine. This direct experience is often described as a profound realization of the interconnectedness of all existence and the presence of God in all things.

Mystical experiences are characterized by a sense of unity, transcendence, and profound insight into the nature of reality. In these states, the boundaries between self and other, subject and object, dissolve, revealing a deeper, more fundamental reality. Mystics often describe their experiences as encounters with the "ground of being" or the "source of all existence," where the divine is experienced as the ultimate reality that underlies all things. Mysticism offers an alternative perspective on existence that emphasizes reality's experiential and transcendent aspects. It challenges the conventional understanding of existence as a material and empirical phenomenon and encourages seekers to explore reality's deeper, spiritual dimensions.

Metaphysical Theories of Reality

Platonism, rooted in the teachings of the ancient Greek philosopher Plato, posits the existence of a realm of abstract, eternal, and unchanging Forms or Ideas, which represent the actual reality beyond the physical world. According to Plato, the physical world is a mere shadow or reflection of the higher reality of the Forms. For example, the Form of Beauty exists independently of any particular beautiful object, and all beautiful things participate in this Form to varying degrees. In Platonism, the material world is seen as impermanent and changeable, while the realm of Forms is eternal and unchanging. The Forms represent the perfect, ideal essence of things, and actual knowledge is attained by contemplating these Forms rather than relying on sensory experience. The highest of all Forms is the Form of the Good, the source of all truth, beauty, and justice.

Platonism has had a lasting influence on Western philosophy, particularly in metaphysics and epistemology. It offers a perspective on existence that emphasizes the existence of a higher, transcendent reality beyond the material world. This perspective challenges us to look beyond the physical and empirical and to seek knowledge of the eternal and unchanging truths that underlie all existence. Panpsychism is a metaphysical theory that posits consciousness as a fundamental and ubiquitous aspect of reality. According to panpsychism, all matter possesses some consciousness or subjective experience, from the smallest particles to the most significant cosmic structures. This view challenges the traditional materialist perspective that consciousness is an emergent property of complex physical systems like the brain.

Panpsychism suggests that consciousness is not limited to humans or animals but is a universal feature of the cosmos. This perspective offers a radical rethinking of the nature of existence, where consciousness is seen as an intrinsic part of the fabric of reality. Panpsychism also provides a potential solution to the hard problem of consciousness, as it posits that consciousness is not something that needs to be "created" by physical processes but is instead a fundamental aspect of all matter. The implications of panpsychism are profound, as it challenges the traditional dichotomy between the mental and the physical and suggests that the universe is inherently conscious. This perspective also raises questions about the nature of the self and the relationship between individual consciousness and the broader, universal consciousness.

Panpsychism has gained attention in recent years as an alternative to both materialism and dualism. It offers a new framework for understanding the nature of consciousness and its role in the cosmos. While still a minority view in philosophy and science, panpsychism has the potential to reshape our understanding of existence and the relationship between mind and matter.

The Question of Ultimate Reality

The question of ultimate reality has been a central theme in metaphysics, with various theories proposing different answers to the nature of existence. Monism is the view that reality is ultimately composed of a single substance or principle, whether material (as in materialism) or spiritual (as in idealism). In contrast, dualism posits that reality consists of two fundamental substances or principles: mind and matter or good and evil. Monistic theories emphasize the unity and interconnectedness of all things, suggesting that the world's diversity expresses a single underlying reality. In contrast, dualistic theories emphasize the distinction and separation between different aspects of reality, such as the physical and the mental or the material and the spiritual. The debate between monism and dualism has profound implications for our understanding of existence, as it raises questions about the nature of the self, the relationship between mind and body, and the ultimate nature of reality. These questions continue to be explored in philosophy and science as thinkers seek to understand the fundamental principles governing existence.

The concept of the infinite and the absolute has been a central theme in metaphysical thought, particularly in the context of ultimate reality. The infinite refers to the idea of boundlessness, whether in space, time, or quantity, while the absolute refers to that which is unconditioned, independent, and self-sufficient. In many metaphysical systems, the infinite and the absolute are identified with the ultimate reality or the divine. For example, in Neoplatonism, the One is the endless source of all existence, from which all things emanate. In Advaita Vedanta, Brahman is the endless and absolute reality that underlies everything. The concept of the infinite challenges our understanding of existence, as it suggests that reality is not limited by the constraints of space, time, or quantity. On the other hand, the absolute challenges the notion of relativity and dependence, suggesting that there is a fundamental reality that is self-sufficient and unchanging.

The infinite and the absolute offer a perspective on existence that emphasizes the transcendence and boundlessness of reality. These concepts challenge us to think beyond the limitations of our finite experience and consider the possibility of an ultimate reality beyond comprehension.

Conclusion

The question of existence is one of the most profound and enduring inquiries in philosophy, science, and religion. It touches on the nature of reality, consciousness, and the self and challenges us to consider the foundations of our understanding of the world. Philosophical perspectives on existence, such as dualism, idealism, materialism, phenomenology, and existentialism, offer different approaches to understanding reality and the nature of the self. Each perspective provides unique insights and raises important questions about the nature of existence and the reliability of our perceptions. Scientific perspectives, including neuroscience, quantum mechanics, and artificial intelligence, have contributed to our understanding of existence by challenging traditional views and opening up new possibilities for exploring the nature of reality. These perspectives raise important questions about the relationship between the mind and the body, consciousness's role, and the universe's nature. Metaphysical and religious perspectives offer alternative frameworks for understanding existence beyond the material and empirical world. These perspectives emphasize reality's spiritual and transcendent aspects and challenge us to consider the possibility of an ultimate reality that underlies all things.

The question of existence cannot be easily answered, and it continues to be a central theme in philosophy, science, and religion. Continued inquiry and exploration are essential for deepening our understanding of reality and the nature of the self. As we continue to explore the nature of existence, we must remain open to new perspectives and willing to challenge our assumptions. The complexity and diversity of the views on existence reflect the richness of human thought and the depth of the mystery of reality. The pursuit of knowledge and understanding is a fundamental aspect of the human experience, and the question of existence invites us to engage with some of the most profound and challenging aspects of our existence.

The question of existence is not merely an

intellectual inquiry but also a source of wonder and mystery. It invites us to reflect on the nature of reality, the meaning of life, and our place in the cosmos. The diversity of perspectives on existence reflects the richness of the human experience and the complexity of our world. Each perspective offers a unique way of understanding reality and invites us to explore the depths of our existence. As we continue to explore the question of existence, we are reminded of the importance of wonder, curiosity, and the pursuit of knowledge. Existence is not just a fact to be explained but a mystery to be embraced and explored.

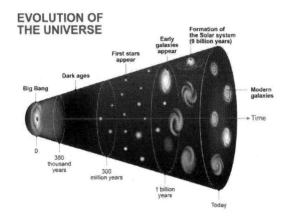

CHAPTER 3: WHAT WAS THERE BEFORE THE BIG BANG, OR WAS THERE A BIG BANG?

The Big Bang theory stands as the prevailing cosmological model that explains the origin and evolution of the universe. According to this theory, the universe began approximately 13.8 billion years ago as a boiling, dense point and has been expanding ever since. This event marks the inception of space, time, matter, and energy, serving as the starting point for the universe as we comprehend it today. However, the question of what existed before the Big Bang, or whether there was a Big Bang, remains a topic of intense debate and speculation among scientists, philosophers, and theologians.

The notion of a "before" the Big Bang challenges our understanding of time, causality, and the universe itself. If the Big Bang signifies the beginning of time, then the concept of "before" may not apply in the traditional sense. Alternatively, some theories suggest that the Big Bang was just one event in a broader, possibly cyclical, cosmic history, introducing the possibility of pre-Big Bang conditions or even a multiverse.

This chapter delves into the Big Bang theory, exploring its implications for our understanding of the universe and examining various hypotheses about what, if anything, existed before this event. We will explore the evidence supporting the Big Bang, alternative cosmological models, and the philosophical and scientific challenges associated with understanding the universe's origins.

The Big Bang Theory and Its Implications

The Big Bang theory is one of modern cosmology's most successful and well-supported models, offering a comprehensive explanation for the observable features of the universe, including its expansion, the cosmic microwave background radiation, and the distribution of galaxies. Understanding the Big Bang theory and its implications is crucial for exploring what might have preceded it.

The Origins of the Big Bang Theory

The development of the Big Bang theory can be traced back to the early 20th century, with contributions from several key figures in physics and astronomy. In 1915, Albert Einstein published his general theory of relativity, which provided a new framework for understanding gravity and the universe's structure. However, Einstein initially believed in a static universe and introduced the cosmological constant into his equations to prevent the universe from collapsing under its gravity.

In the 1920s, Russian physicist Alexander Friedmann and the Belgian priest and astronomer Georges Lemaître independently derived solutions to Einstein's equations describing an expanding universe. Lemaître proposed that the universe began as a "primeval atom" or "cosmic egg," which later expanded and evolved into the universe we observe today. This idea laid the groundwork for the Big Bang theory.

The concept of an expanding universe gained further support in 1929 when American astronomer Edwin Hubble discovered that galaxies were moving away from each other, with more distant galaxies receding at higher velocities. This observation, Hubble's law, provided strong evidence for the universe's expansion and suggested that it must have originated from a hot, dense state.

The term "Big Bang" was coined in 1949 by British astronomer Fred Hoyle, who intended it as a pejorative term while advocating for the alternative steady-state theory. However, the name stuck, and the Big Bang theory continued to gain acceptance as more evidence emerged to support it. In the 1940s and 1950s, scientists such as George Gamow, Ralph Alpher, and Robert Herman developed the concept of primordial nucleosynthesis, predicting that the early universe would have produced light elements such as hydrogen and helium. Their work also led to the prediction of the cosmic microwave background (CMB) radiation, a faint glow that would permeate the universe as a remnant of the Big Bang.

The discovery of the CMB in 1965 by Arno Penzias and Robert Wilson provided substantial confirmation of the Big Bang theory. The CMB represents the thermal radiation left over when the universe became transparent to radiation about 380,000 years after the Big Bang. Its uniformity and temperature are consistent with the predictions of the Big Bang model, providing one of the most compelling pieces of evidence for the theory.

Key Features of the Big Bang

One central tenet of the Big Bang theory is that the universe has expanded since its inception. This expansion is not akin to an explosion in space but rather an expansion of space itself. As space expands, galaxies move away from each other, and the overall density of the universe decreases over time.

The universe's expansion is described by the metric of space-time in general relativity, known as the Friedmann-Lemaître-Robertson-Walker (FLRW) metric. The rate of expansion is characterized by the Hubble constant, which relates the velocity of a galaxy's recession to its distance from us. Observations of distant galaxies and the redshift of their light provide evidence for this ongoing expansion.

The CMB is critical evidence supporting the Big

Bang theory. It is the remnant radiation from when the universe was about 380,000 years old and had cooled enough for electrons and protons to combine into neutral atoms, a process known as recombination. This allowed photons to travel freely through space, creating the CMB. The CMB is remarkably uniform, with slight temperature fluctuations corresponding to the seeds of all future structures in the universe, such as galaxies and clusters.

Another critical prediction of the Big Bang theory is the relative abundance of light elements, such as hydrogen, helium, and lithium, produced during the first few minutes of the universe's existence. This process, known as Big Bang nucleosynthesis, occurred as the universe cooled and allowed nuclear reactions. The observed abundance of these light elements in the universe matches the predictions of the Big Bang model, providing further evidence for the theory. The precise ratios of hydrogen to helium and other light elements are consistent with the conditions expected in the early universe.

Implications for the Nature of Time and Space

The Big Bang theory implies that time began with the Big Bang. According to general relativity, time and space are intertwined in the fabric of space-time, and the Big Bang represents the point at which space-time itself came into existence. Before the Big Bang, there was no time as we understand it, making the concept of "before" the Big Bang problematic from a conventional standpoint. This idea challenges our intuitive understanding of time, which we typically consider linear and infinite. If time began with the Big Bang, then asking what happened before the Big Bang may be akin to asking what lies north of the North Pole—a question that has no meaningful answer within our current understanding of space-time.

The universe's expansion also has profound implications for our understanding of space. Space is not simply a void in which objects exist; it is an active, dynamic entity that can stretch, bend, and expand. The expansion of space is responsible for the observed redshift of distant galaxies and the increasing distance between cosmic objects over time. This expansion challenges our traditional notions of space as a static, unchanging backdrop for events. Instead, space is seen as a dynamic component of the universe that plays a crucial role in shaping the structure and evolution of the cosmos.

Alternative Cosmological Models

While the Big Bang theory is the most widely accepted model of the universe's origin and evolution, it is not the only one. Over the years, several alternative cosmological models have been proposed, each offering different explanations for the universe's origins and challenging the idea that the Big Bang was the definitive beginning of everything.

Steady-State Theory

The steady-state theory, developed by Fred Hoyle, Thomas Gold, and Hermann Bondi in the 1940s, posits that the universe has no beginning or end and has always existed in a constant, unchanging state. According to this theory, the universe is eternally expanding, with new matter continuously created to maintain a constant average density. This process would ensure that the universe appears the same at all times and in all directions, a concept known as the "perfect cosmological principle." The steady-state theory was initially proposed as an alternative to the Big Bang theory and was popular among some astronomers in the mid-20th century. Proponents of the steady-state model argued that it avoided the need for a singular beginning, which they saw as problematic and inconsistent with the principle of uniformity.

However, the steady-state theory faced significant challenges as observational evidence increasingly supported the Big Bang model. The discovery of the CMB in 1965 dealt a major blow to the steady-state theory, as this radiation could not be explained by an eternal, unchanging universe. Additionally, the observed abundance of light elements and the distribution of galaxies were more consistent with the predictions of the Big Bang theory. As a result, the steady-state theory has largely fallen out of favor in the scientific community, although it remains an important historical alternative to the Big Bang model. The decline of the steady-state theory highlights the importance of empirical evidence in shaping our understanding of the universe's origins.

Oscillating Universe Model

The oscillating universe model, also known as cyclic cosmology, proposes that the universe undergoes a series of cycles of expansion and contraction, each beginning with a Big Bang and ending with a Big Crunch. In this model, the universe is not a one-time event but a repeating process with no true beginning or end. According to the oscillating universe model, the Big Bang we observe is just one phase in an eternal cycle of cosmic evolution. After each Big Crunch, the universe "rebounds" in a new Big Bang, forming a new universe.

This process could continue indefinitely, with each cycle lasting billions of years.

The oscillating universe model offers an intriguing alternative to a singular beginning, as it suggests that the universe's history is infinite and cyclical. However, this model faces several challenges, both theoretical and observational. One of the key challenges is the issue of entropy. The universe would accumulate more entropy in each cycle, leading to an eventual "heat death" where no further cycles could occur. This raises questions about whether the universe could truly undergo infinite cycles. Additionally, current observations suggest that the universe's expansion is accelerating, driven by dark energy. If this acceleration continues, it would prevent the universe from ever collapsing back into a Big Crunch, effectively ruling out the oscillating universe model as a viable explanation for the universe's origins.

Inflationary Cosmology

Inflationary cosmology is a modification of the Big Bang theory that addresses several key problems with the original model, such as the horizon problem, the flatness problem, and the monopole problem. Inflation proposes that in the first fraction of a second after the Big Bang, the universe underwent an exponential expansion, growing by a factor of at least 10^{26} in a tiny fraction of a second. This rapid expansion, known as inflation, would have smoothed out any initial irregularities in the universe and set the stage for the more gradual expansion we observe today. Inflation also explains why the universe appears so uniform on large scales and its geometry is so close to flat.

One of the most intriguing aspects of inflationary cosmology is the idea of eternal inflation, which suggests that inflation never ends but continues in different regions of space-time. In this scenario, our observable universe is just one "bubble" in a vast, possibly infinite, multiverse, where various areas experience different physical laws and conditions. Eternal inflation raises the possibility of infinite universes, each with unique properties. This multiverse concept challenges our traditional understanding of the universe as a single, isolated entity and opens up new possibilities for understanding the nature of reality. While eternal inflation and the multiverse remain speculative and complex to test, they provide a fascinating framework for exploring the question of what, if anything, existed before the Big Bang. If our universe is just one of many, the question of what preceded the Big Bang may be reframed as a question about the larger context in which our universe exists.

Quantum Cosmology and the Birth of the Universe

Quantum cosmology applies the principles of quantum mechanics to the study of the universe's origins, suggesting that the universe may have emerged from a quantum fluctuation in a pre-existing quantum field. According to this view, the Big Bang was not the beginning of everything but rather the transition from a quantum state to a classical, expanding universe. In quantum mechanics, particles can spontaneously appear and disappear due to quantum fluctuations, known as "vacuum fluctuations." Some physicists propose that the entire universe could have emerged from such a fluctuation, with the Big Bang representing the point at which the quantum fluctuation "inflated" into the observable universe.

Quantum cosmology also explores the role of quantum gravity, a theoretical framework that seeks to reconcile general relativity with quantum mechanics. Quantum gravity aims to describe space-time behavior at the smallest scales, where the effects of quantum mechanics become significant. One of the critical challenges in quantum cosmology is understanding how space-time emerged from a quantum state. Some theories, such as loop quantum gravity and string theory, suggest that space-time is not a continuous entity but is composed of discrete "quanta" or "loops." These theories propose that the Big Bang was a quantum event where space-time transitioned from a quantum to a classical state.

Quantum cosmology offers a new perspective on the universe's origins, suggesting that the Big Bang may have been a quantum event rather than a singular beginning. This raises the possibility that the universe's origins are deeply connected to the laws of quantum mechanics, challenging our understanding of time, space, and causality.

Philosophical and Theological Considerations

The question of what existed before the Big Bang, or whether there was a Big Bang at all, is not only a scientific inquiry but also a philosophical and theological one. These considerations explore the implications of the Big Bang theory and alternative cosmological models for our understanding of existence, creation, and the nature of reality.

The Concept of Creation Ex Nihilo

The idea of creation ex nihilo, or "creation from nothing," is central in many religious traditions, particularly in the Abrahamic faiths (Judaism, Christianity, and Islam). According to this concept, the universe was created by a divine being or force from nothing, with no pre-existing matter or energy. Some have interpreted the Big Bang theory as a scientific confirmation of this idea, suggesting that the universe had a definite beginning and was brought into existence by a transcendent cause.

Creation ex nihilo raises profound questions about the nature of existence and the relationship between the universe and its creator. If the universe was created from nothing, what does this imply about the nature of the creator and the purpose of creation? Does the existence of a beginning suggest that the universe has a specific purpose or goal, or is it simply the result of a random or arbitrary event? The concept of creation ex nihilo also presents philosophical challenges, particularly regarding causality and the nature of "nothingness." How can something come from nothing? Is it possible for a cause to exist without any pre-existing conditions or materials? These questions challenge our understanding of causality and the nature of existence, as they suggest that the universe's origins may be beyond the scope of human reasoning and comprehension.

Some philosophers argue that the concept of creation ex nihilo is paradoxical and requires rethinking the nature of causality and existence. Others suggest that the idea of creation from nothing may be more of a metaphor or a theological concept rather than a literal description of the universe's origins.

The Nature of Time and Eternity

The question of what existed before the Big Bang is closely tied to the nature of time and eternity. In traditional cosmological models, time is seen as a linear, unidirectional flow, with the Big Bang marking the beginning of time itself. However, some philosophical and theological perspectives suggest that time may not be an absolute, linear construct but rather something relative or nonexistent in the context of the universe's origins. In the context of creation, some theologians and philosophers propose atemporal or timeless creation, where the creator exists outside of time and brings the universe into existence without being bound by temporal constraints. In this view, the creator is eternal and unchanging, existing in a state of timelessness distinct from the temporal world.

Atemporal creation raises essential questions about the nature of time and the relationship between the creator and the created world. How can a timeless being create a temporal universe? What does it mean for time to have a beginning, and how does this relate to the concept of eternity? Unlike the idea of temporal creation, some philosophical perspectives, such as eternalism and the block universe theory, propose that all moments exist simultaneously in a "block" of space-time. In this view, the past, present, and future are all equally real, and the flow of time is an illusion created by our consciousness.

The block universe theory challenges our conventional understanding of time as a linear progression and suggests that the universe's entire history exists as a single, unchanging entity. This perspective raises questions about the nature of causality, free will, and the possibility of change. If all moments exist simultaneously, what does this mean for our experience of the present and our ability to make choices? The concept of eternalism also intersects with theological ideas of eternity and the nature of the divine. If the universe exists as a timeless block, how does this relate to the idea of a timeless creator? Can the concepts of time and eternity be reconciled, or are they fundamentally incompatible?

The Multiverse and the Problem of Fine-Tuning

One of the most intriguing questions in cosmology is the apparent fine-tuning of the universe's physical constants, which appear to be precisely calibrated to allow for the existence of life. The anthropic principle suggests that the universe's properties are finely tuned because we, as conscious observers, exist to observe them. In other words, the universe must have the properties necessary for life because we are here to ask the question. The fine-tuning of the universe has led some to argue for the existence of a multiverse, where multiple universes with different physical constants exist. In this view, our universe is just one of many, and the fine-tuning we observe results from a selection effect —only in universes with the right conditions for life would observers exist to notice the fine-tuning.

The concept of the multiverse raises essential theological and philosophical questions. If there are infinite universes, each with different properties, what does this mean for the idea of a purposeful creation? Does the existence of multiple universes diminish the significance of our universe, or does it suggest a more complex and grand design? Some theologians argue that the multiverse concept is compatible with the idea of divine creation, suggesting that a creator could have brought multiple universes into existence, each with its purpose and design. Others see the multiverse as a challenge to traditional theological concepts, raising questions about the uniqueness of our universe and the nature of divine intention.

The multiverse also intersects with philosophical debates about the nature of reality and the limits of human knowledge. If the multiverse exists, how can we ever know the true nature of reality? What does it mean to exist in one universe among many, and how do we determine the boundaries of our knowledge and understanding?

The Mystery of the Universe's Origins

The question of what existed before the Big Bang, or whether there was a Big Bang, ultimately leads us to the limits of human knowledge and understanding. The universe's origins touch on some of the most profound and challenging questions in cosmology, philosophy, and theology, and, likely, we may never have definitive answers to these questions. The mystery of the universe's origins reminds us of the vastness and complexity of the cosmos and the limitations of our ability to comprehend it. While science has made significant strides in understanding the universe, many unanswered questions and mysteries lie beyond the reach of empirical observation and measurement.

Pursuing knowledge about the universe's origins is a scientific endeavor and a philosophical and spiritual endeavor. It invites us to explore the most profound questions of existence, meaning, and purpose and to confront the possibility that some aspects of reality may remain forever beyond our understanding. The mystery of the universe's origins also invites us to embrace the unknown and approach the question with humility and openness. The quest to understand the universe is a journey of discovery, and the unanswered questions and mysteries are a reminder of the endless possibilities for exploration and learning.

The unknown is not something to be feared but rather an opportunity for growth and discovery. The search for answers to what existed before the Big Bang or whether there was a Big Bang challenges us to think beyond the boundaries of our current knowledge and explore new ideas and perspectives. In the end, the mystery of the universe's origins is a testament to the wonder and beauty of the cosmos. It reminds us that the universe is a place of infinite possibilities and that searching for knowledge is a journey that is never truly complete.

Conclusion

The Big Bang theory is the most widely accepted model of the universe's origin, providing a comprehensive explanation for the observable features of the cosmos, including its expansion, the cosmic microwave background radiation, and the distribution of galaxies. The Big Bang marks the beginning of space, time, matter, and energy, serving as the starting point for the universe as we know it. However, alternative cosmological models, such as the steady-state theory, the oscillating universe model, inflationary cosmology, and quantum cosmology, offer different perspectives on the universe's origins and challenge the idea of a singular beginning. These models explore the possibility of an eternal universe, a cyclic cosmology, a multiverse, or a quantum birth of the universe.

The question of what existed before the Big Bang, or whether there was a Big Bang at all, raises profound philosophical and theological questions about the nature of time, causality, and the universe's origins. Concepts such as creation ex nihilo, atemporal creation, the multiverse, and the fine-tuning of the universe challenge our understanding of existence and the limits of human knowledge. The universe's origins are among the most profound and challenging questions in cosmology, philosophy, and theology. Continued inquiry and exploration are essential for deepening our understanding of the cosmos and the nature of existence. As we continue to explore the universe's origins, we must remain open to new ideas and perspectives and be willing to challenge our assumptions. The complexity and diversity of cosmological models reflect the richness of human thought and the depth of the mystery of the universe.

Pursuing knowledge about the universe's origins is a journey of discovery that invites us to explore the most profound questions of existence, meaning, and purpose. It challenges us to think beyond the boundaries of our current knowledge and embrace the unknown. The question of what existed before the Big Bang, or whether there was a Big Bang, is a scientific inquiry and a source of wonder and inspiration. It invites us to reflect on the nature of reality, the meaning of existence, and our place in the cosmos. The mystery of the universe's origins reminds us of the vastness and complexity of the cosmos and the limitations of our ability to comprehend it. It challenges us to approach the question with humility and openness and to embrace the unknown as an opportunity for growth and discovery. In the end, the search for answers to the questions of the universe's origins is a testament to the wonder and beauty of the cosmos. It reminds us that the universe is a place of infinite possibilities and that searching for knowledge is a journey that is never truly complete.

CHAPTER 4: EINSTEIN UNDERSTOOD

Albert Einstein is widely recognized as one of the most influential scientists of the 20th century. His groundbreaking contributions to physics have profoundly reshaped our understanding of the universe. Through his theories of special relativity general relativity, and his pivotal contributions to quantum mechanics, Einstein revolutionized how we perceive space, time, gravity, and the fundamental nature of reality. These theories advanced scientific knowledge and laid the groundwork for technological innovations that have become integral to our daily lives, from GPS systems to the exploration of black holes and the expanding universe.

Einstein's theories fundamentally challenged the classical Newtonian worldview that had dominated physics for centuries. His work introduced revolutionary concepts, such as the relativity of time and space, the equivalence of mass and energy, and the curvature of space-time caused by gravity. These ideas not only transformed the field of physics but also carried profound philosophical implications, prompting new questions about the nature of reality, the limits of human knowledge, and the relationship between the observer and the observed.

In this chapter, I will explore Einstein's major theories—special relativity, general relativity, and his contributions to quantum mechanics—focusing on their key concepts, the empirical evidence supporting them, and their far-reaching impact on science and society. We will also examine the philosophical implications of Einstein's work and how his theories continue influencing modern physics and our understanding of the universe.

Special Relativity: The Relativity of Space and Time

One of Einstein's most groundbreaking contributions to physics is the theory of special relativity, which he published in 1905. Special relativity fundamentally altered our understanding of space, time, and motion, introducing concepts that are now central to modern physics. The theory challenged the classical Newtonian view of absolute space and time, proposing instead that space and time are relative and interwoven, a notion that has reshaped the landscape of theoretical physics.

The Postulates of Special Relativity

One of the critical postulates of special relativity is the constancy of the speed of light. Einstein proposed that the speed of light in a vacuum is constant and independent of the motion of the observer or the light source. This was a revolutionary idea based on the null results of the Michelson-Morley experiment, which sought to detect the "aether," a supposed medium through which light waves were thought to travel. The experiment failed to show any variation in the speed of light due to Earth's motion through space, concluding that the speed of light is a universal constant.

The constancy of the speed of light has profound implications for our understanding of space and time. It implies that the laws of physics, including the speed of light, are the same for all observers, regardless of their state of motion. This challenges the classical notion of absolute time and space, suggesting that these concepts are not fixed but relative to the observer's frame of reference.

Another critical concept in special relativity is the relativity of simultaneity, which states that two events that are simultaneous in one frame of reference may not be simultaneous in another frame moving relative to the first. This idea challenges the classical notion of an absolute, universal time that flows uniformly for all observers. Instead, special relativity implies that time is not a universal constant but is experienced differently depending on the observer's state of motion.

Time Dilation and Length Contraction

One of the most famous predictions of special relativity is time dilation, where time appears to pass more slowly for an observer moving at high speeds relative to a stationary observer. This effect becomes significant at speeds close to the speed of light. For instance, an astronaut traveling at near-light speed would experience time much more slowly than people on Earth. Upon returning, the astronaut would find that less time has passed for them than for those who remained on Earth.

Time dilation has been experimentally confirmed in numerous ways, including observations of particles moving at high speeds and precise measurements of time using atomic clocks on fast-moving aircraft and satellites. These experiments demonstrate that time dilation is not just a theoretical concept but a real effect with practical implications. Time dilation also has important implications for our understanding of aging and the limits of space travel, as illustrated by the famous "twin paradox."

Another consequence of special relativity is length contraction, where the length of an object moving at high speeds appears contracted or shortened along the direction of its motion, as observed from a stationary frame of reference. Like time dilation, length contraction becomes significant at speeds close to the speed of light. This effect has been observed in high-energy particle experiments, where particles moving at near-light speeds travel shorter distances than expected based on their rest length. Length contraction further illustrates the relativity of space and time in special relativity.

Mass-Energy Equivalence: $E=mc^2$

Perhaps the most well-known and significant result of special relativity is the equation $E=mc^2$, which expresses the equivalence of mass and energy. According to this equation, mass can be converted into energy and vice versa, with the conversion factor being the speed of light squared. This means that even a small amount of mass can be converted into a large amount of energy, as seen in nuclear reactions and particle-antiparticle annihilation.

The equation $E=mc^2$ has had profound implications for both science and technology. It explains the tremendous energy released in nuclear reactions, such as those that power the Sun and nuclear weapons. It also underlies the principles of nuclear power and the potential for harnessing fusion energy as a sustainable energy source.

Mass-energy equivalence has also significantly impacted our understanding of the universe. It suggests that mass and energy are not separate entities but are interchangeable under certain conditions. This idea has been confirmed through numerous experiments involving particle collisions and the observation of energy-mass conversions in astrophysical phenomena. The equation $E=mc^2$ provides a framework for understanding how energy and mass interact in the context of space-time and gravity, setting the stage for Einstein's later work on general relativity.

General Relativity: The Curvature of Space-Time and Gravity

Einstein's theory of general relativity, published in 1915, extended the principles of special relativity to include gravity, providing a new understanding of how gravity works and affects the fabric of space-time. General relativity replaced Newton's theory of gravity with a more comprehensive framework that describes gravity not as a force between masses but as the curvature of space-time caused by mass and energy.

The Equivalence Principle

One of the key ideas that led Einstein to develop general relativity was the equivalence principle, which states that the effects of gravity are locally indistinguishable from the effects of acceleration. In other words, an observer in a closed room cannot tell whether they are experiencing gravity or being accelerated in space. The equivalence principle suggests that gravity is not a force in the traditional sense but rather a manifestation of the curvature of space-time.

This principle implies that space-time is not a static, unchanging entity but is influenced by the presence of mass and energy. The curvature of space-time caused by mass and energy is what we perceive as gravity, and objects move along curved paths in this curved space-time. This idea challenges the classical Newtonian view of gravity as a force acting at a distance between two masses.

The Geometry of Space-Time

In general relativity, gravity is described by the curvature of space-time, which is determined by the distribution of mass and energy. The mathematical framework for this curvature is provided by the Einstein field equations, a set of ten interrelated differential equations describing how space-time's curvature is related to the energy and momentum of matter and radiation. These equations are highly complex and challenging to solve, but they provide a powerful tool for understanding the behavior of massive objects and the universe's structure.

One of the most famous solutions to the Einstein field equations is the Schwarzschild solution, which describes the space-time geometry around a spherically symmetric, non-rotating mass, such as a star or a black hole. This solution provides the basis for understanding the behavior of black holes and the bending of light by gravity, known as gravitational lensing.

General relativity also predicts that time passes more slowly in strong gravitational fields, an effect known as gravitational time dilation. This effect is most pronounced near massive objects, such as black holes, where the curvature of space-time is extreme. Near the event horizon of a black hole, time appears to slow down to an outside observer, effectively coming to a standstill at the event horizon.

Black holes are one of the most dramatic predictions of general relativity, representing regions of space-time where the curvature becomes so extreme that not even light can escape. The study of black holes has provided valuable insights into the nature of gravity, the behavior of matter under extreme conditions, and the limits of our understanding of the universe.

The Bending of Light and Gravitational Lensing

One of the key predictions of general relativity is that light is deflected, or bent, by the presence of a massive object due to the curvature of space-time. This effect, known as gravitational lensing, was first confirmed during a solar eclipse in 1919 when astronomers observed the deflection of starlight passing near the Sun. This observation provided one of the first experimental confirmations of general relativity and established Einstein as a leading figure in science.

Gravitational lensing occurs because light follows the curvature of space-time, and in the presence of a massive object, this curvature causes the light to bend. The amount of deflection depends on the mass of the object and the distance of the light path from the object. Gravitational lensing can produce various optical effects, including multiple images of a distant object, arcs, and rings, known as Einstein rings.

Gravitational lensing has become a powerful tool in modern astrophysics and cosmology. It allows astronomers to study mass distribution in galaxies and galaxy clusters, including the distribution of dark matter, which does not emit light but can be detected through its gravitational effects. Lensing also enables the observation of distant galaxies and quasars that would otherwise be too faint to detect. Gravitational lensing has provided valuable insights into the universe's large-scale structure, the distribution of dark matter, and the behavior of light in curved space-time.

Cosmological Implications of General Relativity

General relativity has profoundly impacted our understanding of the universe's structure and evolution. One of the critical implications of general relativity is the prediction of an expanding universe, which was first proposed by Georges Lemaître and later confirmed by Edwin Hubble's observations of distant galaxies. The Friedmann-Lemaître-Robertson-Walker (FLRW) metric, a solution to the Einstein field equations, describes a homogeneous, isotropic universe and forms the basis of the Big Bang theory and the standard model of cosmology.

The cosmological constant, initially introduced by Einstein to maintain a static universe, has taken on new significance in modern cosmology as the term associated with dark energy, the mysterious force driving the universe's accelerated expansion. The cosmological constant represents the energy density of space, or vacuum energy, and plays a crucial role in understanding the universe's fate.

General relativity also predicts the existence of black holes, regions of space-time where gravity is so strong that nothing, not even light, can escape. Black holes have become a central research topic in modern astrophysics, with observations of stellar-mass black holes, supermassive black holes at the centers of galaxies, and the detection of gravitational waves from black hole mergers. The study of black holes has provided valuable insights into gravity's nature, matter's behavior under extreme conditions, and the limits of our understanding of space-time.

The universe's fate is closely tied to black holes' behavior and space-time expansion. If the universe's expansion continues indefinitely, it could lead to a "Big Freeze," where galaxies drift apart, stars burn out, and the universe becomes cold and dark. Alternatively, if the universe were to collapse in a "Big Crunch," black holes could play a central role, with space-time eventually contracting into a singularity. These scenarios highlight the profound implications of general relativity for our understanding of the universe's ultimate fate.

Einstein and Quantum Mechanics: The Struggle with Uncertainty

While Einstein is best known for his work on

relativity, he also made significant contributions to the development of quantum mechanics, a branch of physics that describes the behavior of particles on the smallest scales. However, Einstein's relationship with quantum mechanics was complex and often contentious, as he struggled with the theory's inherent uncertainty and probabilistic nature.

The Photoelectric Effect and the Quantum Nature of Light

In 1905, the same year he published his theory of special relativity, Einstein also published a groundbreaking paper on the photoelectric effect, for which he later received the Nobel Prize in Physics in 1921. In this paper, Einstein proposed that light is quantized and consists of discrete packets of energy called photons. This idea was a radical departure from the classical wave theory of light and provided strong evidence for the quantum nature of light.

The photoelectric effect occurs when light shines on a material and ejects electrons from its surface. According to classical physics, the energy of the ejected electrons should depend on the intensity of the light. Still, experiments showed that the power of the electrons relies only on the frequency of the light, not its intensity. Einstein explained this by proposing that light is composed of photons, each with energy proportional to its frequency, and that only photons with sufficient energy could eject electrons.

Einstein's work on the photoelectric effect provided crucial support for the emerging theory of quantum mechanics. It helped establish the concept of wave-particle duality, where light exhibits wave-like and particle-like properties depending on the experimental conditions.

The Uncertainty Principle and Einstein's Objections

One of the fundamental principles of quantum mechanics is the Heisenberg uncertainty principle, which states that it is impossible to simultaneously know a particle's exact position and momentum with arbitrary precision. The more accurately one of these quantities is known, the less accurately the other can be determined. This principle challenges the classical notion of determinism, where the future behavior of a system can be predicted with certainty, given complete knowledge of its initial conditions.

The uncertainty principle has profound implications for our understanding of reality and the limits of human knowledge. It suggests that the behavior of particles is inherently probabilistic and that certain aspects of a system are fundamentally indeterminate. This principle is a cornerstone of quantum mechanics and has been confirmed through numerous experiments, including the double-slit experiment and the behavior of particles in quantum systems.

Despite his contributions to quantum mechanics, Einstein was deeply troubled by the theory's probabilistic nature and its implications for determinism and causality. He famously expressed his discomfort with the theory by stating, "God does not play dice with the universe," indicating his belief that the underlying reality of the universe must be deterministic and governed by precise laws. Einstein's objections to quantum mechanics were rooted in his belief that the theory was incomplete and failed to capture reality's true nature. He argued that the probabilistic nature of quantum mechanics reflected our ignorance of the underlying variables, which he called "hidden variables."

Einstein's critique of quantum mechanics played a crucial role in the development of the theory, particularly in the exploration of quantum entanglement and the concept of non-locality. His work on the EPR (Einstein-Podolsky-Rosen) paradox, which questioned the completeness of quantum mechanics, laid the groundwork for later developments in quantum theory, including the concept of quantum entanglement.

The EPR Paradox and Quantum Entanglement

In 1935, Einstein and Boris Podolsky and Nathan Rosen published a paper known as the EPR paradox, which challenged the completeness of quantum mechanics. The EPR paradox focused on quantum entanglement, where two particles become correlated so that one particle's state instantaneously affects the other's state, regardless of the distance between them. This phenomenon, known as non-locality, seemed to violate the principle of locality, which states that objects can only be influenced by their immediate surroundings.

The EPR paradox questioned whether quantum mechanics could completely describe reality, suggesting that the strange behavior of entangled particles indicated the existence of hidden variables that quantum mechanics did not account for. However, the EPR paradox sparked decades of debate and experimental research, culminating in the work of physicist John Bell, who formulated Bell's theorem in 1964. Bell's theorem provided a way to test whether hidden variables could explain the predictions of quantum mechanics or whether they genuinely required non-locality.

Experiments testing Bell's theorem, most notably those conducted by physicist Alain Aspect and his collaborators in the 1980s, confirmed the predictions of quantum mechanics and demonstrated that quantum entanglement is a natural phenomenon. These experiments showed that any local hidden variable theory could not explain the correlations between entangled particles, effectively confirming the non-local nature of quantum mechanics.

The confirmation of quantum entanglement has had profound implications for our understanding of reality and has led to the development of new technologies, such as quantum cryptography and quantum computing. It also challenges our classical intuitions about space, time, and causality, suggesting that the universe is more interconnected and mysterious than previously thought.

Einstein's Legacy in Quantum Mechanics

Despite his discomfort with quantum mechanics, Einstein's critique of the theory played a crucial role in its development. His work on the EPR paradox and his insistence on questioning the completeness of quantum mechanics stimulated important debates and led to the exploration of new ideas and concepts like quantum entanglement and non-locality.

Einstein's legacy in quantum mechanics is not one of rejection but of constructive skepticism. His willingness to challenge the prevailing views of the time and to ask deep, probing questions about the nature of reality has had a lasting impact on the field. His work continues to inspire physicists to explore the mysteries of the quantum world and to seek a deeper understanding of the universe.

The ongoing quest for a unified theory that would reconcile general relativity with quantum mechanics remains one of the most critical challenges in modern physics. Einstein spent the later years of his life searching for such a theory, which he believed would provide a comprehensive understanding of all the fundamental forces of nature and restore determinism and causality to the description of reality. Although he was unsuccessful in this quest, his work laid the groundwork for the ongoing search for a "Theory of Everything" that continues today.

Philosophical Implications of Einstein's Theories

Einstein's theories of relativity and his contributions to quantum mechanics have had profound philosophical implications, challenging our understanding of space, time, causality, and the nature of reality. These implications have sparked essential debates in the philosophy of science and have influenced a wide range of philosophical discussions.

The Nature of Space and Time

One of the critical philosophical implications of Einstein's theories is the block universe concept, which arises from the theory of relativity. In the block universe, all moments—past, present, and future—exist simultaneously in a four-dimensional space-time continuum. This challenges the classical view of time as a linear progression from past to future and suggests that the flow of time is an illusion created by our consciousness.

The block universe raises essential questions about the nature of time, causality, and free will. If all moments exist simultaneously, what does this mean for our experience of the present and our ability to make choices? Are we truly free to shape our future, or is everything predetermined by the structure of space-time?

The block universe also challenges our understanding of the self and personal identity. If the self exists as a series of moments in space-time, how do we account for the continuity of consciousness and the experience of being a unified, enduring individual? These questions continue to be explored in the philosophy of time and metaphysics, with Einstein's theories as a central point of reference.

Causality and the Nature of Reality

Einstein's work on quantum mechanics and his critique of the theory's probabilistic nature have had significant philosophical implications for our understanding of causality and reality. Quantum mechanics challenges the classical notion of determinism, where the future behavior of a system can be predicted with certainty, given complete knowledge of its initial conditions. Instead, quantum mechanics suggests that certain aspects of reality are fundamentally indeterminate and governed by probabilities.

This challenges our understanding of causality and raises important questions about the nature of reality. If the behavior of particles is inherently probabilistic, what does this mean for the concept of cause and effect? How do we reconcile the indeterminacy of quantum mechanics with our everyday experience of a stable, predictable world?

The philosophical implications of quantum mechanics have been the subject of intense debate, with different interpretations of the theory offering different answers to these questions. The Copenhagen interpretation, which is the most widely accepted, embraces the indeterminacy and probabilistic nature of quantum mechanics. In contrast, the many-worlds interpretation suggests that all possible outcomes of a quantum event occur in parallel universes, preserving determinism but at the cost of introducing an infinite number of realities.

The Relationship Between the Observer and the Observed

Relativity and quantum mechanics challenge the classical view of the observer as a passive entity that records the world's state. In relativity, the observer's frame of reference plays a crucial role in determining the measurements of space and time, leading to effects such as time dilation and length contraction. This suggests that the observer is an active participant in shaping the description of reality.

In quantum mechanics, the observer plays an even more central role, as observation can influence the behavior of quantum particles, as seen in the observer effect and the collapse of the wave function. This challenges the classical notion of an objective reality that exists independently of observation and raises important questions about the nature of measurement and the role of consciousness in the quantum world.

The relationship between the observer and

the observed has profound philosophical implications, particularly in the philosophy of science and metaphysics. It challenges the traditional distinction between subject and object and suggests that our descriptions of reality are shaped by our interactions with the world. This has led to new perspectives on the nature of reality, knowledge, and the limits of human understanding.

Einstein's Legacy in Philosophy and Science

Einstein's theories have had a lasting impact on the philosophy of science, particularly in discussions of the nature of scientific theories, the role of observation and measurement, and the limits of scientific knowledge. His work has inspired new approaches to understanding the relationship between theory and experiment. It has raised important questions about the nature of scientific realism and the interpretation of physical theories.

Einstein's critique of quantum mechanics has also significantly influenced the philosophy of science, particularly in debates about the interpretation of quantum mechanics, the nature of causality, and the role of determinism in science. His work continues to be a central point of reference in discussions of the nature of reality and the foundations of physics.

Einstein's legacy continues to shape modern physics, with his theories of relativity and his contributions to quantum mechanics serving as the foundation for much of contemporary research. The quest to reconcile general relativity with quantum mechanics, the search for a unified theory, and the exploration of the nature of space-time and gravity are all active research that builds on Einstein's work.

Einstein's theories have also profoundly impacted our understanding of the universe, from the behavior of black holes and the universe's expansion to the development of new technologies, such as GPS and gravitational wave detectors. His work continues to inspire new generations of physicists, philosophers, and thinkers to explore the universe's mysteries and seek a deeper understanding of the nature of reality.

Conclusion

Albert Einstein's contributions to physics, particularly his theories of special relativity and general relativity, and his work on quantum mechanics have profoundly shaped our understanding of the universe. His theories introduced new concepts, such as the relativity of space and time, the curvature of space-time caused by gravity, and the equivalence of mass and energy.

Special relativity challenged the classical Newtonian view of absolute space and time, introducing concepts such as time dilation, length contraction, and mass-energy equivalence. Numerous experiments have confirmed these ideas and have had significant implications for our understanding of the universe and the developing of new technologies.

General relativity provided a new understanding of gravity as the curvature of space-time caused by mass and energy. The theory has been confirmed through observations of gravitational time dilation, gravitational lensing, and the behavior of black holes, and it has had a profound impact on our understanding of the structure and evolution of the universe.

Einstein's relationship with quantum mechanics

was complex, as he struggled with the theory's probabilistic nature and its implications for determinism and causality. His critique of quantum mechanics played a crucial role in the development of the theory, particularly in the exploration of quantum entanglement and the concept of non-locality.

Einstein's theories have had profound philosophical implications, challenging our understanding of space, time, causality, and the nature of reality. His work has influenced a wide range of philosophical discussions and continues to be a central point of reference in the philosophy of science.

The Importance of Continued Inquiry and Exploration

Einstein's work reminds us of the importance of questioning assumptions, challenging prevailing views, and seeking a deeper understanding of the universe. His willingness to explore new ideas and push the boundaries of human knowledge has had a lasting impact on both science and philosophy.

The ongoing quest to reconcile general relativity with quantum mechanics, the search for a unified theory, and the exploration of the nature of space-time and gravity are all active research that builds on Einstein's work. These areas of inquiry continue to push the boundaries of our understanding and hold the potential for discoveries and insights.

Einstein's Legacy as a Source of Wonder and Inspiration

Einstein's contributions to physics and his philosophical insights inspire new generations of thinkers and researchers. His work challenges us to explore the mysteries of the universe, seek a deeper understanding of reality, and embrace the cosmos' wonder and beauty.

The legacy of Einstein's work is not just in the theories and equations he developed but in the spirit of inquiry and exploration he embodied. His willingness to challenge the status quo and explore new ideas is a model for future generations of scientists, philosophers, and thinkers.

In the end, Einstein's work reminds us that the universe is a place of infinite possibilities, and the search for knowledge is a journey that is never truly complete. His legacy continues to shape our understanding of the universe and inspire us to seek a deeper understanding of the nature of reality.

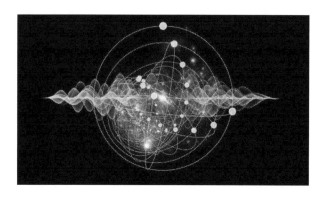

CHAPTER 5: QUANTUM MECHANICS FOR LAYMEN

Quantum mechanics is one of the most successful yet perplexing theories in the history of science. This branch of physics deals with the behavior of matter and energy on the smallest scales, at the level of atoms and subatomic particles. Unlike classical mechanics, which describes the motion of everyday objects, quantum mechanics operates in a realm where the rules are entirely different, often defying our common-sense understanding of the world. The theory has been pivotal in explaining phenomena that classical physics could not account for and has led to the development of numerous technologies integral to modern life, such as semiconductors, lasers, and magnetic resonance imaging (MRI).

The development of quantum mechanics in the early 20th century marked a fundamental shift in our understanding of nature. It introduced concepts such as wave-particle duality, uncertainty, superposition, and entanglement, which challenge our classical notions of reality and have profound implications for our understanding of the universe. While quantum mechanics is mathematically complex, its core ideas can be appreciated without delving into the intricacies of its equations.

This chapter explores the basic principles of quantum mechanics in a way accessible to laypeople. I will examine the historical context in which quantum mechanics emerged, explore its key concepts, and discuss the philosophical implications of quantum mechanics and how it challenges our understanding of reality. Finally, we will look at some of the most famous experiments and phenomena in quantum mechanics and consider how this theory has influenced science and technology.

The Birth of Quantum Mechanics: A Historical Overview

Quantum mechanics did not emerge fully formed but resulted from discoveries and developments over several decades. The late 19th and early 20th centuries were periods of significant scientific advancement, during which many of the classical theories of physics were called into question by new experimental evidence.

One of the first indications that classical physics was incomplete came from studying blackbody radiation. A blackbody is an idealized object that absorbs all radiation falling on it and emits radiation based on its temperature. Classical physics, particularly the laws of thermodynamics and electromagnetism, could not accurately predict the distribution of energy emitted by a blackbody at different wavelengths. According to classical theory, the energy radiated at ultraviolet frequencies should increase without bound, leading to what became known as the "ultraviolet catastrophe."

This problem was resolved by Max Planck in 1900, who proposed that energy is not emitted continuously but in discrete packets, or "quanta." This radical departure from the classical view laid the groundwork for quantum theory, though it was not fully understood or accepted.

Another significant challenge to classical physics came from studying the photoelectric effect, where light shining on a metal surface causes the ejection of electrons. Classical wave theory predicted that the energy of the ejected electrons should depend on the intensity of the light, but experiments showed that it relies only on the light's frequency. In 1905, Albert Einstein explained this by proposing that light consists of discrete packets of energy called photons, each with energy proportional to its frequency. This explanation, built on Planck's idea of quantized energy, provided strong evidence for the particle nature of light and earned Einstein the Nobel Prize.

The study of atomic spectra also revealed the limitations of classical physics. When excited, atoms emit light at specific wavelengths, creating a spectrum of lines unique to each element. Classical physics could not explain why these lines appeared at particular wavelengths. In 1913, Niels Bohr proposed a model of the atom where electrons orbit the nucleus in discrete energy levels. According to Bohr's model, electrons can jump between these levels, emitting or absorbing energy in discrete amounts corresponding to the observed spectral lines. This model successfully explained the spectral lines of hydrogen and introduced the concept of quantized energy levels in atoms. However, it still had limitations later addressed by the full development of quantum mechanics.

The Development of Quantum Theory

In 1924, Louis de Broglie proposed that particles such as electrons also exhibit wave-like properties, a hypothesis later confirmed by experiments. De Broglie's idea of matter waves provided a new perspective on particle nature and led to wave mechanics' development. The concept of wave-particle duality became a central theme in quantum mechanics, where entities such as electrons and photons are understood to behave as both particles and waves depending on the context.

In 1927, Werner Heisenberg formulated the uncertainty principle, which states that it is impossible to simultaneously know a particle's exact position and momentum with arbitrary precision. The more accurately one is known, the less accurately the other can be determined. This principle is a direct consequence of the wave-particle duality of matter and has profound implications for our understanding of reality.

Erwin Schrödinger further developed wave

mechanics by formulating the Schrödinger equation, a fundamental quantum mechanics equation that describes how a physical system's quantum state changes over time. The Schrödinger equation treats particles as wave functions, representing the probability distribution of a particle's position and momentum.

Max Born introduced the probabilistic interpretation of the wave function, where the square of the wave function's amplitude gives the probability density of finding a particle in a particular position. This interpretation, known as the Born rule, is a crucial feature of quantum mechanics and reflects the inherently probabilistic nature of the theory.

The Copenhagen Interpretation

The Copenhagen interpretation, developed primarily by Niels Bohr and Werner Heisenberg, is the most widely accepted interpretation of quantum mechanics. According to this interpretation, the wave function represents the complete description of a quantum system, and its evolution is governed by the Schrödinger equation. However, when a measurement is made, the wave function collapses to a single eigenstate corresponding to the observed outcome.

The concept of wave function collapse is one of the most debated aspects of the Copenhagen interpretation. It suggests that the act of measurement plays a fundamental role in determining the outcome of a quantum system, leading to questions about the role of the observer and the nature of reality.

The Copenhagen interpretation also introduces the principle of complementarity, which states that different experimental setups can reveal different aspects of a quantum system. Still, these aspects cannot be observed simultaneously. For example, an experiment designed to measure the wave-like properties of a particle will not reveal its particle-like properties, and vice versa.

Niels Bohr, one of the founding figures of quantum mechanics, emphasized the importance of embracing the paradoxes and contradictions of quantum theory. He argued that quantum mechanics requires a new way of thinking about the world, where classical concepts such as determinism and objective reality must be reconsidered.

Key Concepts of Quantum Mechanics

Quantum mechanics introduces several key concepts that challenge our classical understanding of reality. These concepts are central to the theory and have profound implications for how we think about the nature of matter, energy, and the universe.

Wave-particle duality is one of the most fundamental concepts in quantum mechanics, describing how particles such as photons and electrons exhibit wave-like and particle-like properties depending on the experimental context. This duality challenges the classical distinction between waves and particles and suggests that the behavior of quantum entities cannot be fully understood using classical concepts.

The double-slit experiment is one of the most famous demonstrations of wave-particle duality. In this experiment, a beam of light or electrons is directed at a barrier with two slits, and the resulting pattern is observed on a screen behind the barrier. When both slits are open, the particles create an interference pattern on the screen, characteristic of wave-like behavior. However, when detectors are placed at the slits to observe which slit the particles pass through, the interference pattern disappears, and the particles behave like individual particles.

The principle of superposition is another key concept in quantum mechanics, stating that a quantum system can exist in multiple states simultaneously until it is measured. For example, an electron in an atom can exist in a superposition of different energy levels, or a photon can exist in a superposition of different polarization states.

Schrödinger's cat is a famous thought experiment proposed by Erwin Schrödinger to illustrate the superposition and wave function collapse paradoxes. In this thought experiment, a cat is placed in a sealed box with a radioactive atom, a Geiger counter, and a vial of poison. If the atom decays, the Geiger counter triggers the release of the poison, killing the cat. If the atom does not decay, the cat remains alive. According to quantum mechanics, the atom exists in a superposition of decayed and undecayed states. Therefore, the cat exists in a superposition of alive and dead states until the box is opened and the system is observed.

Quantum entanglement is a phenomenon where two or more particles become correlated so that the state of one particle instantaneously affects the state of the other, regardless of the distance between them. This phenomenon, which Albert Einstein called "spooky action at a distance," challenges the classical concept of locality, where objects can only be influenced by their immediate surroundings.

The uncertainty principle, formulated by Werner

Heisenberg, states that certain pairs of physical properties, such as position and momentum, cannot be simultaneously measured with arbitrary precision. The more accurately one property is known, the less accurately the other can be determined.

Famous Experiments and Phenomena in Quantum Mechanics

Quantum mechanics has been confirmed through numerous experiments that demonstrate quantum systems' strange and counterintuitive behavior. These experiments have provided valuable insights into the nature of matter and energy and have led to the development of new technologies that rely on quantum principles.

The double-slit experiment is one of the most famous demonstrations of wave-particle duality and the role of the observer in quantum mechanics. When particles such as electrons or photons pass through two slits and are not observed, they create an interference pattern on a screen, characteristic of wave-like behavior. However, when detectors are placed at the slits to observe which slit the particles pass through, the interference pattern disappears, and the particles behave like individual particles.

The EPR paradox, proposed by Albert Einstein, Boris Podolsky, and Nathan Rosen, questioned whether quantum mechanics could completely describe reality. The paradox focused on quantum entanglement, where one particle's state instantaneously affects another's state, regardless of the distance between them. The EPR paradox suggested that the strange behavior of entangled particles indicated the existence of hidden variables that quantum mechanics did not account for.

According to classical physics, Quantum tunneling is a phenomenon where particles can pass through energy barriers that they would not be able to cross. In classical physics, if a particle does not have enough energy to overcome a barrier, it will be reflected. However, in quantum mechanics, there is a finite probability that the particle will "tunnel" through the barrier and appear on the other side.

Quantum superposition is the phenomenon where a quantum system can exist in multiple states simultaneously until it is measured. This leads to interference effects, where the different states of the system can combine to produce patterns that have no classical analog.

The Philosophical Implications of Quantum Mechanics

Quantum mechanics has had profound philosophical implications, challenging our understanding of reality, causality, and the nature of the universe. These implications have sparked essential debates in the philosophy of science and have influenced a wide range of philosophical discussions.

One of the most profound implications of quantum mechanics is its challenge to the classical notion of an objective reality that exists independently of observation and measurement. The role of the observer in quantum mechanics, particularly in the concept of wave function collapse, suggests that observation plays a fundamental role in determining the outcome of a quantum system.

The many-worlds interpretation of quantum

mechanics offers a different perspective on the nature of reality. According to this interpretation, all possible outcomes of a quantum event occur in parallel universes, each representing a distinct branch of reality.

Quantum mechanics challenges the classical notion of determinism, where the future behavior of a system can be predicted with certainty, given complete knowledge of its initial conditions. The uncertainty principle and the probabilistic nature of quantum mechanics suggest that certain aspects of a quantum system are inherently indeterminate and that probabilities rather than deterministic laws govern the behavior of particles.

The role of the observer in quantum mechanics, particularly in the concept of wave function collapse, has profound implications for our understanding of reality and the nature of measurement. The observer effect suggests that observation plays a fundamental role in determining the outcome of a quantum system, challenging the classical distinction between subject and object.

Quantum mechanics has raised important questions about the limits of human knowledge and the possibility of a complete description of reality. The uncertainty principle and the probabilistic nature of quantum mechanics suggest that certain aspects of reality are fundamentally indeterminate and that there may be limits to what we know about the world.

Quantum Mechanics and Modern Technology

Quantum mechanics has not only transformed our understanding of the universe but has also led to the development of new technologies that have revolutionized modern life. These technologies rely on the principles of quantum mechanics and demonstrate the practical applications of this theory.

Semiconductors and transistors are the building blocks of modern electronics, and their development was made possible by the principles of quantum mechanics. Semiconductors have electrical conductivity between a conductor and an insulator, and their behavior can be controlled by doping them with impurities. Transistors are semiconductor devices that can amplify or switch electronic signals, and they are used in a wide range of electronic devices, from computers to smartphones.

Lasers, which amplify light to produce a highly coherent and focused beam, rely on the principles of quantum mechanics. The laser operation is based on stimulated emission, where an incoming photon promotes an electron in an excited state to emit a second photon of the same frequency and phase. The principles of quantum mechanics govern this process and allow for light amplification.

Magnetic resonance imaging (MRI) is a medical imaging technique that relies on the principles of quantum mechanics to produce detailed images of the body's internal structures. MRI works by detecting the nuclear spin of hydrogen atoms in the body, which are aligned by a strong magnetic field and then perturbed by a radiofrequency pulse. The resulting signals are detected and processed to create images of the body's tissues.

Quantum computing is an emerging field that seeks to harness the principles of quantum mechanics to perform calculations beyond the reach of classical computers. Quantum computers use qubits, which can exist in a superposition of states, allowing them to perform multiple calculations simultaneously.

Quantum communication systems use the principles of quantum mechanics, particularly quantum entanglement and superposition, to transmit information securely. Quantum cryptography, for example, uses entangled photons to create secure communication channels immune to eavesdropping.

Conclusion

Quantum mechanics is a fundamental theory of physics that describes the behavior of matter and energy on the smallest scales. It introduces concepts such as wave-particle duality, superposition, entanglement, and the uncertainty principle, which challenge our classical understanding of reality.

The development of quantum mechanics was driven by a series of discoveries and experiments that revealed the limitations of classical physics. Key figures in the development of quantum mechanics include Max Planck, Albert Einstein, Niels Bohr, Werner Heisenberg, Erwin Schrödinger, and Louis de Broglie.

Quantum mechanics has profound philosophical implications, challenging our understanding of reality, causality, and the nature of the universe. It raises crucial questions about the role of the observer, the nature of measurement, and the limits of human knowledge.

Quantum mechanics has led to the development of new technologies that have revolutionized modern life, including semiconductors, lasers, magnetic resonance imaging (MRI), quantum computing, and quantum communication systems. These technologies demonstrate the practical applications of quantum mechanics and its impact on science and technology.

Quantum mechanics remains an active area of research, with ongoing efforts to understand its implications and develop new technologies based on its principles. The quest to reconcile quantum mechanics with general relativity, the search for a unified theory, and the exploration of quantum computing and communication are all areas of active inquiry.

Quantum mechanics is not just a scientific theory but a source of wonder and inspiration. It challenges our classical intuitions and invites us to explore the mysteries of the universe at the most minor scales. The strange and counterintuitive behavior of quantum systems reveals a more complex and interconnected world than we ever imagined.

Studying quantum mechanics encourages us to embrace the unknown and approach the universe's mysteries with curiosity and open-mindedness. It reminds us that pursuing knowledge is a never-complete journey and that the universe is a place of infinite possibilities.

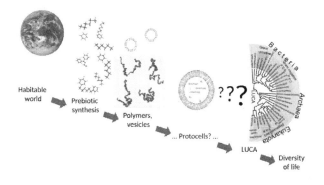

CHAPTER 6: FROM ENERGY TO LIVING ENTITY

The transition from energy to life represents one of the most profound and mysterious processes in the universe. It is a journey from simple molecules and energy to complex, self-replicating entities capable of metabolism, adaptation, and evolution. Understanding how the basic building blocks of the universe coalesced to form life intersects biology, chemistry, and physics and touches on broader inquiries into the origins of life and consciousness.

From the primordial conditions of early Earth to the sophisticated machinery within cells, this transition involves forming complex organic molecules, self-organization, information processing, and the delicate balance of energy that sustains life. This chapter explores the scientific understanding of how energy drives the transformation from non-living matter to living entities. I will examine the role of thermodynamics in biological systems, the principles of self-organization, the chemistry of life's building blocks, and the origins of the first living cells. Additionally, we will discuss the philosophical implications of this transition, particularly concerning the nature of life, emergence, and the boundary between the living and the non-living.

The Role of Energy in Biological Systems

Energy is the driving force behind all physical processes in the universe, including those that give rise to life. In biological systems, energy is not merely a requirement for sustaining life; it plays a fundamental role in the processes leading to life's formation. Understanding how energy influences these processes is crucial for comprehending the emergence of life from non-living matter.

The Laws of Thermodynamics and Life

The First Law of Thermodynamics, also known as energy conservation, states that energy cannot be created or destroyed, only converted from one form to another. This principle is evident in biological systems through processes like metabolism, where chemical energy stored in molecules such as glucose is converted into usable forms like adenosine triphosphate (ATP) to power cellular functions.

The Second Law of Thermodynamics states that a closed system's total entropy (disorder) will always increase over time. This principle might seem to contradict the highly ordered nature of living organisms, but it does not when considering that living systems are not closed. They constantly exchange energy and matter with their surroundings, allowing them to maintain and even increase their internal order at the expense of increasing the entropy of their environment.

In biological systems, the concept of free energy, specifically Gibbs free energy, is vital in understanding how biochemical reactions occur. Gibbs free energy (G) represents the portion of a system's energy that can perform work at a constant temperature and pressure. For a reaction to occur spontaneously, the change in Gibbs free energy (ΔG) must be negative, meaning the system releases free energy and moves toward a more stable state.

Energy Conversion in Biological Systems

Photosynthesis is the process by which plants, algae, and certain bacteria convert light energy from the sun into chemical energy stored in glucose and other organic molecules. It is the primary means by which energy from the sun is captured and made available to life on Earth. Photosynthesis occurs in two main stages: the light-dependent reactions, which generate ATP and NADPH, and the Calvin cycle, which uses these molecules to convert carbon dioxide into glucose.

Cellular respiration is the process by which cells extract energy from organic molecules like glucose to produce ATP. This process involves glycolysis, the citric acid cycle, and oxidative phosphorylation. Glycolysis breaks down glucose into pyruvate, producing ATP and NADH. Pyruvate is further processed in the citric acid cycle, releasing carbon dioxide and generating more

NADH and FADH2. These electron carriers donate their electrons to the electron transport chain, where redox reactions create a proton gradient that drives ATP synthesis.

Fermentation is an anaerobic process that allows cells to produce ATP without oxygen. It is critical for many organisms, especially those in oxygen-poor environments. In fermentation, glycolysis produces ATP by breaking down glucose into pyruvate, which is then converted into lactate, ethanol, and carbon dioxide, depending on the organism. This process regenerates NAD+, allowing glycolysis to continue.

The Chemistry of Life: Building Blocks and Complex Molecules

Life is based on complex chemistry involving interactions between various molecules and compounds. These chemical processes are essential for forming, maintaining, and reproducing living organisms.

The Elements of Life

Carbon is the fundamental element in the chemistry of life due to its ability to form four covalent bonds, allowing for the creation of complex and stable molecules such as carbohydrates, lipids, proteins, and nucleic acids. Hydrogen, oxygen, nitrogen, phosphorus, and sulfur are also critical elements in life's chemistry, contributing to biological molecules' structure and function.

The Formation of Biomolecules

Carbohydrates serve as primary energy sources and structural components for living organisms. Simple sugars like glucose can link to form complex carbohydrates, such as starch, glycogen, and cellulose, essential for energy storage and structural support.

Lipids are a diverse group of hydrophobic molecules crucial for energy storage, membrane structure, and signaling. Fats and oils, known as triglycerides, serve as long-term energy storage molecules, while phospholipids form cell membranes' structural basis.

Proteins are complex molecules composed of amino acids that perform various functions, including catalyzing biochemical reactions, providing structural support, transporting molecules, and regulating cellular processes. The structure of a protein, determined by the sequence of amino acids, is critical for its function.

Nucleic acids, including DNA and RNA, store and transmit genetic information. DNA carries the genetic blueprint for development and reproduction, while RNA plays various roles in expressing genetic information.

The Emergence of Complex Molecules and Life

Prebiotic chemistry refers to the chemical processes on early Earth before life emerged, leading to forming the basic building blocks of life, such as amino acids, nucleotides, and simple sugars. The Miller-Urey experiment demonstrated that organic molecules could form under prebiotic conditions, providing insights into life's origins.

The RNA world hypothesis suggests that RNA was the first self-replicating molecule with its ability to store genetic information and catalyze reactions. Over time, RNA-based systems evolved in complexity, leading to the development of DNA and proteins as the primary genetic material and catalysts, respectively.

Protocells, simple membrane-bound structures, represent an intermediate stage between non-living matter and living cells. They likely formed from the spontaneous assembly of lipids and other molecules, providing a protected environment for RNA replication and other biochemical processes.

Self-Organization and the Emergence of Life

Self-organization is a fundamental principle in the transition from energy to living entities, where simple components spontaneously arrange into more complex structures. This process is closely related to the concept of emergence, where complex behavior arises from simpler components' interactions.

Non-equilibrium thermodynamics studies systems that are not equilibrium and are driven by external energy inputs. Biological systems maintain order and complexity far from thermodynamic equilibrium through metabolic processes.

Feedback loops, mechanisms where the output of a process influences the process itself, play a critical role in maintaining homeostasis and regulating cellular activities. Autocatalysis, a type of positive feedback loop, likely played a key role in the early stages of life.

The formation of cellular membranes, the evolution of metabolism, and the transition from RNA-based systems to DNA and protein-based life were crucial steps in the origin of the first cells.

Philosophical Implications of the Transition from Energy to Life

The transition from energy to life raises profound philosophical questions about the nature of life, emergence, and the boundary between the living and the non-living.

Defining life is a central philosophical question, with life typically characterized by properties such as metabolism, reproduction, and adaptation. However, some borderline cases, such as viruses, challenge traditional definitions.

Emergence is central to understanding how life arises from non-living matter, suggesting that life is an emergent property resulting from simpler components' interactions.

Viruses challenge the traditional definition of life, as they possess some characteristics of life but lack others. Advances in synthetic biology and the creation of artificial life forms have further blurred the boundary between the living and the non-living.

The study of the transition from energy to life also has implications for astrobiology and the search for extraterrestrial life, raising questions about the nature of life and its potential forms beyond Earth.

Conclusion

The transition from energy to life is a complex process involving energy, chemistry, and self-organization. It is a crucial area of inquiry with profound implications for understanding life and its origins.

Energy plays a central role in biological systems, driving the processes that lead to the formation and maintenance of life. The chemistry of life involves forming complex molecules from simple building blocks, and self-organization and emergence are critical principles in this transition.

The origin of the first cells involved forming cellular membranes, evolving metabolism, and transitioning from RNA-based systems to DNA and protein-based life.

The transition from energy to life raises philosophical questions about the nature of life, emergence, and the boundary between the living and the non-living. These questions challenge our understanding of life and its origins, with important implications for astrobiology, synthetic biology, and the search for extraterrestrial life.

Continued inquiry and exploration are essential for deepening our understanding of the processes that lead to the emergence of life. This research has the potential to answer fundamental questions about the nature of life and its place in the universe.

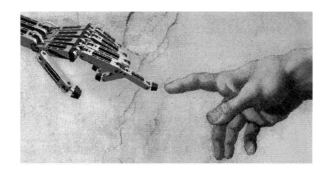

CHAPTER 7: QUANTUM ENTANGLEMENT WITHIN THE BRAIN

Quantum entanglement is one of the most enigmatic phenomena in quantum mechanics, where two or more particles become linked so that the state of one particle is instantaneously correlated with the state of another, regardless of the distance between them. Albert Einstein famously described this phenomenon as "spooky action at a distance," which challenges our classical understanding of space, time, and causality.

While quantum entanglement is well-established in subatomic particles, its potential relevance to biological systems, particularly the human brain, is a subject of growing interest and debate. Could quantum entanglement influence brain function, consciousness, cognition, or mental processes? This question lies at the intersection of physics, neuroscience, and philosophy, holding profound implications for our understanding of the mind and its relationship to the physical world.

This chapter explores the concept of quantum entanglement, its foundational principles, and its experimental verification. I will then delve into the emerging field of quantum biology, examining evidence and hypotheses suggesting that quantum phenomena, including entanglement, might influence biological processes. Finally, we will focus on the brain, discussing various theories that propose a connection between quantum entanglement and consciousness, cognition, and neural processing. We will also address the challenges and criticisms of these theories and the philosophical implications of a quantum brain.

The Basics of Quantum Entanglement

To understand the potential role of quantum entanglement in the brain, it is essential first to grasp the basic principles of quantum entanglement and its significance in quantum mechanics. Entanglement is a quantum phenomenon that classical physics cannot explain and has been experimentally confirmed in numerous studies.

The Principles of Quantum Entanglement

At the heart of quantum mechanics is the concept of superposition, where a quantum system can exist in multiple states simultaneously until it is observed or measured. The wave function is a mathematical representation of this superposition, describing the probabilities of finding the system in each possible state. When two or more particles become entangled, their wave functions become intertwined, meaning that the state of one particle cannot be described independently of the other(s). This entanglement leads to correlations between the particles that persist regardless of the distance between them.

One of the most puzzling aspects of quantum entanglement is its non-locality, where entangled particles exhibit correlations that appear to be instantaneous, even when separated by large distances. This seems to violate the classical idea that information cannot travel faster than the speed of light. However, experiments have confirmed non-locality, most notably those testing Bell's theorem. These experiments have shown that any local hidden variable theory cannot explain the correlations between entangled particles.

John Bell's theorem, formulated in 1964, provided a way to test whether the predictions of quantum mechanics, specifically those related to entanglement, could be explained by classical physics or hidden variables. Bell's theorem demonstrated that specific predictions of quantum mechanics are incompatible with the principle of local realism, which states that local factors determine the properties of particles and that information cannot travel faster than light. Experiments testing Bell's theorem, such as those conducted by Alain Aspect and his collaborators in the 1980s, have confirmed the non-local nature of quantum entanglement.

The Implications of Quantum Entanglement

Quantum entanglement challenges many classical concepts, such as locality, causality, and the independence of distant objects. In classical physics, objects are assumed to have well-defined properties independent of observation, and the speed of light limits any influence between objects. However, entanglement suggests that particles can be instantaneously connected, even when separated by vast distances, challenging our classical intuitions about the nature of reality.

Beyond its theoretical implications, quantum entanglement has practical applications in emerging technologies, such as quantum computing, cryptography, and teleportation. In quantum computing, entangled qubits can perform complex calculations more efficiently than classical computers. In quantum cryptography, entanglement can create secure communication channels immune to eavesdropping. In quantum teleportation, entanglement enables the transfer of quantum states between particles without physically moving them.

Quantum Biology: Exploring Quantum Phenomena in Biological Systems

Quantum biology is an emerging field that explores the possibility that quantum phenomena, such as entanglement, superposition, and tunneling, play a role in biological processes. While biology has traditionally been understood through the lens of classical physics and chemistry, recent research suggests that quantum mechanics may be relevant to understanding certain biological functions, particularly those involving complex, highly sensitive, and efficient processes.

Evidence of Quantum Phenomena in Biology

One of the most well-studied examples of

quantum phenomena in biology is photosynthesis, the process by which plants, algae, and certain bacteria convert light energy into chemical energy. Recent research has shown that quantum coherence, where particles exist in a superposition of states, plays a role in energy transfer efficiency in photosynthetic complexes. Light-absorbing molecules known as chromophores in photosynthesis capture photons and transfer the absorbed energy through a series of molecular complexes to the reaction center. Studies using ultrafast spectroscopy have revealed that the energy transfer in these complexes occurs through quantum coherence, allowing the energy to explore multiple pathways simultaneously and find the most efficient route to the reaction center.

Another area where quantum phenomena have been proposed to play a role is in the magnetic sense of certain animals, such as birds, fish, and insects. This sense, known as magnetoreception, allows these animals to navigate using the Earth's magnetic field. One of the leading hypotheses is that magnetoreception is mediated by a quantum process involving entangled electron pairs in a protein called cryptochrome. According to this hypothesis, when cryptochrome is exposed to light, it generates a pair of entangled electrons. The Earth's magnetic field influences the spin states of these electrons, affecting the chemical reactions within the cryptochrome and ultimately impacting the animal's ability to sense the magnetic field.

Enzymes, which are biological catalysts, accelerate chemical reactions in living organisms. Recent research suggests that quantum tunneling, where particles pass through energy barriers that they cannot overcome classically, might play a role in certain enzymatic reactions. In enzyme catalysis, tunneling could allow reactants to convert into products more efficiently by bypassing the energy barriers associated with the response, potentially explaining the high catalytic efficiency of certain enzymes.

Challenges and Criticisms of Quantum Biology

One of the main challenges in applying quantum mechanics to biological systems is the issue of decoherence, where quantum systems lose their coherence and behave classically due to interactions with their environment. Maintaining quantum coherence long enough to influence biological functions is challenging in biological systems, as they are typically warm, wet, and noisy. Critics argue that the decoherence timescales in biological systems are too short for quantum phenomena to play a significant role in most biological processes.

Another criticism of quantum biology is the relative lack of direct experimental evidence for quantum effects in most biological systems. While studies on photosynthesis, magnetoreception, and enzyme catalysis provide intriguing hints, the field is still in its early stages. Experimental challenges include isolating and measuring quantum effects in complex living systems and the need for advanced techniques to study quantum phenomena in biological contexts.

Quantum Entanglement in the Brain: Theoretical Perspectives

The idea that quantum entanglement might play a role in brain function has gained attention in recent years, particularly in theories attempting to explain consciousness, cognition, and neural processing. While these theories remain speculative and controversial, they offer intriguing possibilities for understanding the brain in a new light.

The Penrose-Hameroff Orchestrated Objective Reduction (Orch-OR) Theory

One of the most well-known theories linking quantum mechanics to brain function is the Orch-OR theory, proposed by physicist Sir Roger Penrose and anesthesiologist Stuart Hameroff. This theory suggests that quantum processes, specifically quantum entanglement and coherence, occur within microtubules, which are structural components of the cytoskeleton in neurons. Microtubules are hollow, cylindrical structures composed of tubulin proteins, and they play a role in maintaining cell shape, facilitating intracellular transport, and organizing cellular division.

The Orch-OR theory posits that consciousness arises from quantum computations involving the entanglement of tubulin proteins within microtubules. These quantum computations are thought to occur in the brain's neurons, influencing synaptic activity and contributing to the emergence of conscious experience. According to the theory, quantum states within microtubules undergo orchestrated objective reduction (OR), which links quantum mechanics with general relativity, leading to conscious experience.

The Quantum Brain Hypothesis

The quantum brain hypothesis is a broader concept that explores the possibility that the brain processes information using quantum principles, such as superposition, entanglement, and tunneling. Proponents of this hypothesis suggest that quantum processes could enhance the brain's computational power, allowing for more efficient information processing, decision-making, and problem-solving. Some researchers have proposed that quantum entanglement could play a role in neural communication, enabling faster and more efficient transfer of information between neurons.

Quantum tunneling has also been proposed as a potential mechanism in the brain, particularly in processes involving ion transfer across cell membranes or the propagation of action potentials along axons. While this idea remains speculative, it suggests tunneling could provide a mechanism for ultra-fast and efficient signaling in the brain.

Quantum Consciousness and the Hard Problem

The "hard problem" of consciousness concerns the nature of subjective experience and how it arises from physical processes in the brain. While much progress has been made in understanding the neural correlates of consciousness, the question of how and why specific brain processes give rise to conscious experience remains unresolved. Some theorists have proposed that quantum mechanics might provide a new framework for addressing this problem, particularly quantum processes' non-local, holistic, and non-deterministic nature.

One aspect of consciousness that has intrigued researchers is the unity of conscious experience, where different sensory inputs and cognitive processes are integrated into a coherent, unified perception of the world. Some have speculated that quantum entanglement could play a role in this integration, allowing for the instantaneous coordination of neural processes across different brain regions.

Criticisms and Challenges of Quantum Brain Theories

Theories proposing a connection between quantum mechanics and brain function, particularly those involving quantum entanglement, have been met with significant skepticism and criticism. These challenges highlight the difficulties of applying quantum mechanics to complex biological systems and the need for rigorous experimental evidence.

The Problem of Decoherence in the Brain

One of the main challenges in applying quantum mechanics to the brain is the issue of decoherence. The brain's warm, wet, and noisy environment makes it difficult for quantum coherence to be maintained over the timescales required for cognitive processes. Critics argue that any quantum states in the brain would rapidly decohere due to interactions with the surrounding environment, leading to classical behavior. This problem of decoherence suggests that quantum effects might be too short-lived to influence neural processes significantly.

The Lack of Empirical Evidence

Another criticism of quantum brain theories is the lack of direct empirical evidence supporting the involvement of quantum processes in brain function. While there are intriguing theoretical proposals and some indirect evidence, such as the role of quantum coherence in photosynthesis, there is little concrete evidence that quantum entanglement or other quantum phenomena play a significant role in the brain. Developing experimental techniques to test for quantum effects in neural processes directly will be essential to advance quantum brain theories.

Philosophical Implications of Quantum Entanglement in the Brain

The possibility that quantum entanglement might play a role in brain function has profound philosophical implications, particularly concerning the nature of consciousness, free will, and the relationship between mind and matter. These implications challenge traditional views in philosophy and neuroscience and open up new avenues for exploring the mysteries of the mind.

The Mind-Body Problem and Quantum Mechanics

The mind-body problem concerns the relationship between the mind (or consciousness) and the physical body (or brain). Traditionally, this problem has been approached from two main perspectives: dualism, which posits that the mind and body are distinct entities, and materialism, which holds that the mind is a product of physical processes in the brain. Quantum theories of mind, particularly those involving quantum entanglement, offer a potential third perspective that blurs the line between dualism and materialism. These theories suggest that consciousness might emerge from quantum processes deeply intertwined with the brain's physical structure, challenging the traditional distinction between mind and matter.

Quantum Entanglement and Free Will

The question of free will is closely related to the nature of determinism and indeterminism in physical theories. In classical physics, the universe is often viewed as deterministic, meaning that the future is entirely determined by the present, with no room for free will. However, quantum mechanics introduces an element of indeterminism, where certain events, such as the collapse of a wave function, are inherently probabilistic and not entirely determined by prior states. Some theorists have suggested that quantum mechanics might provide a basis for free will, particularly the indeterminism associated with quantum entanglement and wave function collapse.

The Holistic Nature of Quantum Consciousness

One of the most intriguing implications of quantum mechanics, particularly quantum entanglement, is the idea of non-locality, where entangled particles exhibit instantaneous correlations regardless of distance. Some theorists have speculated that non-locality might provide a framework for understanding the unity of conscious experience, where different sensory inputs and cognitive processes are integrated into a coherent, unified perception of the world. This holistic view challenges the traditional, reductionist approach to neuroscience, where the brain is often viewed as a collection of independent modules or networks.

Conclusion

Summary of Key Points:

- Quantum entanglement is a fundamental phenomenon in quantum mechanics, where particles become instantaneously correlated regardless of distance, challenging classical concepts of locality, causality, and independence.

- Quantum biology explores the possibility that quantum phenomena, such as entanglement, coherence, and tunneling, play a role in biological processes. Evidence from studies on photosynthesis, magnetoreception, and enzyme catalysis suggests that quantum mechanics may be relevant to understanding certain biological functions.

- Theories proposing a connection between quantum mechanics and brain function, such as the Orch-OR theory and the quantum brain hypothesis, suggest that quantum entanglement and other quantum processes might play a role in consciousness, cognition, and neural processing. These theories are speculative and controversial but offer intriguing possibilities for understanding the brain in a new light.

- Quantum brain theories face significant challenges, particularly the problem of decoherence in the brain's warm and noisy environment and the lack of direct empirical evidence for quantum effects in neural processes. Addressing these challenges will be crucial for advancing the field and providing more definitive evidence for the role of quantum mechanics in the brain.

The Importance of Continued Inquiry and Exploration:

- The study of quantum entanglement in the brain is an emerging and speculative area of research that challenges our understanding of both quantum mechanics and neuroscience. Continued inquiry and exploration are essential for advancing our knowledge of this topic and testing the validity of quantum brain theories.

- Advances in experimental techniques, such as quantum-enhanced sensors and ultrafast spectroscopy, will be crucial for detecting and measuring quantum effects in the brain. These technologies could provide new insights into the potential role of quantum mechanics in neural processes and contribute to a deeper understanding of consciousness and cognition.

Quantum Entanglement in the Brain as a Source of Wonder and Inspiration:

- The idea that quantum entanglement might play a role in the brain's functioning is a source of wonder and inspiration, offering new perspectives on the nature of consciousness and the mysteries of the mind. It challenges traditional views in philosophy and science and invites us to explore the universe's potential for interconnectedness and holism.

- The study of quantum entanglement in the brain also raises profound philosophical questions about the nature of reality, the possibility of free will, and the relationship between mind and matter. It challenges us to reconsider our assumptions about the brain and consciousness and to explore new frameworks for understanding the mind in a quantum universe.

- Whether or not quantum entanglement plays a significant role in the brain, exploring this idea highlights the importance of interdisciplinary research and the potential for quantum mechanics to revolutionize our understanding of biology, neuroscience, and consciousness. It reminds us that the mysteries of the mind are far from fully understood and that the search for knowledge is an ongoing journey that pushes the boundaries of science and philosophy.

CHAPTER 8: DO WE NEED TO AGE?

Aging is a universal and inevitable process that affects all living organisms, characterized by a gradual decline in physiological function, increased vulnerability to diseases, and, ultimately, death. Traditionally viewed as an unavoidable aspect of life, aging has come under renewed scrutiny as advances in biology, medicine, and biotechnology raise the possibility of slowing, halting, or even reversing the aging process.

The question "Do we need to age?" is scientific, deeply philosophical, and ethical. It challenges our understanding of life, mortality, and the natural order. What would be the implications for individuals, societies, and the planet if aging could be prevented or significantly delayed? How would it affect our concepts of identity, purpose, and the human experience?

In this chapter, we will explore the biological basis of aging, examining various theories and mechanisms that contribute to the aging process. I will discuss genetic, cellular, and molecular factors involved in aging and how they interact to produce observable signs of aging. Additionally, we will delve into the latest research on anti-aging interventions, such as caloric restriction, senolytics, telomere extension, and genetic engineering. Finally, we will consider the philosophical and ethical implications of extending the human lifespan, addressing questions about the desirability and feasibility of achieving "biological immortality."

The Biology of Aging: Mechanisms and Theories

Aging is a multifactorial and complex process involving numerous biological pathways and mechanisms. Understanding the biology of aging requires exploring various theories proposed to explain why and how aging occurs, along with the critical cellular and molecular processes driving the aging process.

Theories of Aging

The **programmed aging theory** posits that aging is a genetically determined process, where an organism's genetic makeup predetermines its physiological decline. According to this theory, aging is an evolutionary adaptation that regulates the lifespan of a species, ensuring individuals do not outlive their reproductive usefulness and allowing for generational turnover. The discovery of "longevity genes," such as those involved in the insulin/IGF-1 signaling pathway in *Caenorhabditis* *elegans*, supports the notion that aging is partly under genetic control. However, critics argue that aging is more likely a result of stochastic processes and cumulative damage rather than a predetermined genetic program.

The **damage accumulation theory**, also known as the "wear and tear" theory, suggests that aging results from the gradual accumulation of damage to cells, tissues, and organs over time. This damage can be caused by various factors, including oxidative stress, DNA mutations, protein misfolding, and environmental insults. Over time, the accumulated damage impairs cellular function, leading to a decline in physiological function and the onset of age-related diseases. The **free radical theory of aging**, proposed by Denham Harman, highlights oxidative stress as a significant contributor to aging.

The **mitochondrial theory of aging** focuses on the role of mitochondria, the cell's energy-producing organelles, in the aging process. Mitochondria generate ATP through oxidative phosphorylation, producing reactive oxygen species (ROS) as byproducts. Over time, these ROS can damage mitochondrial DNA (mtDNA), leading to a decline in mitochondrial function and increased ROS production, creating a vicious cycle of damage. Mitochondrial dysfunction is associated with various age-related diseases, such as neurodegenerative disorders and cardiovascular disease.

The **telomere shortening theory** posits that the gradual loss of telomere length is a crucial driver of aging and age-related diseases. Telomeres are repetitive DNA sequences at the ends of chromosomes that protect them from degradation. With each round of cell division, telomeres shorten, eventually leading to cellular senescence or apoptosis when they become critically short. Studies have shown that shorter telomeres are associated with an increased risk of age-related diseases and a shorter lifespan. Conversely, interventions that maintain telomere length, such as telomerase activation, have extended lifespan in animal models.

Key Cellular and Molecular Processes in Aging

Cellular senescence is a state of permanent cell cycle arrest that occurs when cells experience stress, such as DNA damage, oxidative stress, or telomere shortening. Senescent cells no longer divide but remain metabolically active, secreting various pro-inflammatory factors, growth factors, and proteases, collectively known as the senescence-associated secretory phenotype (SASP). While senescence is a protective mechanism to prevent the proliferation of damaged cells, the accumulation of senescent cells over time contributes to tissue dysfunction and chronic inflammation, known as "inflammaging".

DNA damage and repair are central to the aging process. DNA damage can arise from oxidative stress, environmental toxins, radiation, and errors during DNA replication. Cells have evolved various DNA repair mechanisms to maintain genomic integrity, but these mechanisms become less efficient with age, accumulating DNA damage. The decline in DNA repair capacity is associated with an increased risk of cancer and other age-related diseases.

Epigenetic changes refer to heritable changes in gene expression that do not involve alterations in the underlying DNA sequence. With age, the epigenetic landscape of cells becomes increasingly dysregulated, leading to changes in gene expression that contribute to aging and age-related diseases. Alterations in DNA methylation patterns are particularly well-studied and can be used to predict biological age through "epigenetic clocks".

Stem cell exhaustion refers to the decline in stem cell function with age, reducing tissue regeneration capacity and increasing the risk of age-related diseases. This decline is driven by factors such as DNA damage, telomere shortening, epigenetic changes, and the accumulation of senescent cells. Stem cell exhaustion is particularly relevant in tissues with high regenerative capacity, such as the skin, blood, and gut.

Proteostasis or protein homeostasis is essential for cellular function, as misfolded or damaged proteins can aggregate and form toxic structures that contribute to age-related diseases, such as Alzheimer's and Parkinson's. With age, the cellular mechanisms responsible for proteostasis, including the ubiquitin-proteasome system and autophagy, become less efficient, accumulating damaged proteins and forming protein aggregates.

Interventions to Slow or Reverse Aging

The growing understanding of the biology of aging has led to the development of various interventions aimed at slowing, halting, or even reversing the aging process. These interventions target critical mechanisms of aging and have shown promise in extending lifespan and improving healthspan in animal models and, in some cases, humans.

Caloric Restriction and Caloric Restriction

Mimetics

Caloric restriction (CR), the practice of reducing calorie intake without malnutrition, has been shown to extend lifespan and delay the onset of age-related diseases in various organisms, from yeast to mammals. The effects of CR on aging are thought to be mediated by multiple mechanisms, including reduced oxidative stress, enhanced DNA repair, improved proteostasis, and increased autophagy. CR influences critical signaling pathways involved in aging, such as the insulin/IGF-1 signaling pathway, the mTOR pathway, and the sirtuin pathway.

Given the challenges of long-term CR in humans, researchers have sought to develop **caloric restriction mimetics (CRMs)**. These compounds mimic the beneficial effects of CR without the need for reduced calorie intake. CRMs target the same signaling pathways as CR and have shown promise in extending lifespan and improving healthspan in animal models. Resveratrol, rapamycin, and metformin are among the most well-known CRMs.

Senolytics: Targeting Senescent Cells

Senolytics are drugs that selectively eliminate senescent cells, which have stopped dividing but remain metabolically active and contribute to chronic inflammation and tissue dysfunction. The accumulation of senescent cells with age is a significant driver of aging and age-related diseases. Removing these cells has been shown to extend lifespan and improve healthspan in animal models.

Studies on senolytics, such as the combination of dasatinib and quercetin, have shown potential in early human trials, improving physical function and reducing senescence markers in elderly individuals. However, using senolytics in humans raises challenges, including the potential for off-target effects and the need for precise dosing and timing.

Telomere Extension and Telomerase Activation

Telomeres, the protective caps at the ends of chromosomes, shorten with each round of cell division, eventually leading to cellular senescence or apoptosis. **Telomerase** is an enzyme that can extend telomeres by adding repetitive DNA sequences to the ends of chromosomes. While most somatic cells have low or absent telomerase activity, interventions that maintain or extend telomere length have shown promise in delaying aging and preventing age-related diseases.

Telomerase activators, such as TA-65, have been shown to increase telomerase activity, extend telomeres in human cells, and improve markers of aging in animal models. However, telomerase activation carries potential risks, particularly the risk of promoting cancer, as cancer cells often rely on telomerase to maintain their telomeres and enable uncontrolled proliferation.

Genetic Engineering and Rejuvenation Therapies

Advances in genetic engineering have opened up new possibilities for targeting the genetic and molecular pathways involved in aging. **Gene therapy** involves introducing, removing, or modifying genes within an individual's cells to treat or prevent disease. Gene therapy has been explored in aging to extend lifespan, enhance health span, and reverse age-related decline.

Rejuvenation therapies aim to restore youthful

function to aged tissues and organs, potentially reversing the effects of aging. These therapies may involve a combination of approaches, including stem cell therapy, gene therapy, and regenerative medicine techniques. Young blood or plasma used to rejuvenate aged tissues and stem cell therapy to replace or repair damaged tissues are among the most promising areas of rejuvenation research.

Philosophical and Ethical Implications of Extending Human Lifespan

The possibility of extending the human lifespan through anti-aging interventions raises profound philosophical and ethical questions. These questions touch on fundamental aspects of human existence, including the nature of life and death, the meaning of aging, and the impact of lifespan extension on individuals, societies, and the planet.

The Nature of Aging and Mortality

Aging has traditionally been seen as a natural and inevitable part of life, with death as the ultimate endpoint. From a biological perspective, aging results from a gradual decline in physiological function driven by genetic, environmental, and stochastic factors. From a philosophical perspective, aging and death are often viewed as integral to the human experience, shaping our understanding of identity, purpose, and the passage of time.

The possibility of extending lifespan or achieving "biological immortality" challenges these traditional views, raising questions about whether aging should be considered a disease that requires treatment or a natural process that should be accepted. Some argue that aging is a fundamental aspect of life that gives meaning to our existence and that attempts to eliminate aging could have unintended consequences.

The Impact of Lifespan Extension on Society

Extending the human lifespan could have far-reaching social and economic implications, affecting healthcare, retirement, population growth, and resource allocation. While increasing lifespan and health span could reduce the burden of age-related diseases and improve the quality of life for older individuals, it could also lead to challenges such as overpopulation, increased demand for healthcare and social services, and potential economic inequality.

The social and economic impact of lifespan extension will depend on various factors, including the accessibility and affordability of anti-aging interventions, the ability of societies to adapt to longer lifespans, and the potential for generational tensions or conflicts. Policymakers, ethicists, and researchers must consider these factors to explore the feasibility and desirability of extending the human lifespan.

The Concept of Biological Immortality

Biological immortality refers to the concept of an organism that does not experience the effects of aging and can potentially live indefinitely, barring injury or disease. While no organism is genuinely biologically immortal, certain species, such as the jellyfish *Turritopsis dohrnii*, exhibit remarkable regenerative abilities and can revert to a juvenile state, effectively escaping the aging process.

The feasibility of achieving biological immortality remains uncertain, as the biological processes underlying aging are complex and not fully understood. While significant progress has been made in extending lifespan and health span in model organisms, translating these findings to humans presents substantial challenges. The potential risks and unintended consequences of pursuing biological immortality must be carefully considered.

Conclusion

Summary of Key Points:

- Aging is a complex and multifactorial process driven by genetic, cellular, and molecular factors. Various theories of aging, such as the programmed aging theory, damage accumulation theory, mitochondrial theory of aging, and telomere shortening theory, provide insights into the mechanisms underlying the aging process.

- Critical cellular and molecular processes involved in aging include cellular senescence, DNA damage and repair, epigenetic changes, stem cell exhaustion, and proteostasis. These processes contribute to the decline in physiological function and the onset of age-related diseases.

- Advances in anti-aging research have led to the development of various interventions aimed at slowing, halting, or reversing the aging process. These interventions include caloric restriction and caloric restriction mimetics, senolytics, telomere extension and telomerase activation, and genetic engineering and rejuvenation therapies.

- The possibility of extending human lifespan raises profound philosophical and ethical questions, including the nature of aging and mortality, the impact of lifespan extension on society, and the feasibility and desirability of achieving biological immortality.

The Importance of Continued Inquiry and Exploration:

- The study of aging and lifespan extension is an ongoing area of research that continues to challenge our understanding of biology, medicine, and ethics. Continued inquiry and exploration are essential for advancing our knowledge of aging and developing safe and effective interventions to promote healthy aging.

- Advances in biotechnology, genetics, and regenerative medicine offer new possibilities for extending human healthspan and lifespan. However, these technologies also raise important ethical and social considerations that must be addressed as we explore the potential for delaying or reversing the aging process.

Aging and Lifespan Extension as a Source of Wonder and Inspiration:

- The study of aging and lifespan extension is a source of wonder and inspiration, offering new perspectives on the nature of life, death, and the human experience. It challenges us to consider the possibilities and limitations of human biology and to explore the potential for extending the boundaries of life.

- The pursuit of lifespan extension invites us to reflect on the value of life, the meaning of aging, and the impact of longevity on individuals and societies. It encourages us to consider the ethical and philosophical implications of extending life and to explore new frameworks for understanding the relationship between biology, aging, and the human experience.

- Whether or not significant lifespan extension is achievable, exploring aging and longevity highlights the importance of promoting health and well-being. It reminds us that the quest for longevity is not just about adding years to life but about improving the quality of life and ensuring that individuals can live healthy, fulfilling lives at all stages of life.

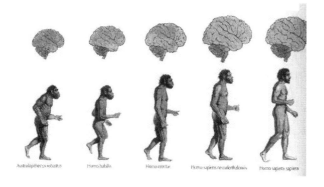

CHAPTER 9: WHAT WOULD WE TALK ABOUT IN THE YEAR 3024?

Imagining life in the year 3024 is a thought experiment that stretches the limits of our understanding and imagination. A thousand years from now, the technological, social, and environmental landscape could be so transformed that it may bear little resemblance to today's world. The changes that could occur over the next millennium might redefine what it means to be human, how we interact with each other, and our place in the universe.

In this chapter, we will explore potential topics

of conversation in the year 3024, considering the advances in technology, society, and culture that could shape the future. I will discuss the impact of artificial intelligence (AI) and machine learning, the exploration and colonization of space, the evolution of human biology through genetic engineering and biotechnology, the potential for achieving immortality, and the future of energy and the environment. I will also examine these developments' social and ethical implications and the philosophical questions they might raise.

This exploration is more than just speculative; it reflects on the long-term consequences of our present actions. The decisions we make today could profoundly influence the world of tomorrow, shaping the lives of future generations and the trajectory of human civilization.

The Future of Technology: Artificial Intelligence and Beyond

Technology has driven human progress, and its rapid advancement over the past few centuries suggests that it will continue to be a central force in shaping the future. By 3024, technology might have evolved to a point where it transcends our current understanding, leading to a world where the boundaries between humans and machines are increasingly blurred.

The Rise of Superintelligent AI

Artificial intelligence (AI) is in its early stages, with most applications falling under "narrow AI"—systems designed to perform specific tasks like image recognition, natural language processing, or autonomous driving. However, the development of artificial general intelligence (AGI), capable of performing any intellectual task a human can do, could become a reality by the 22nd or 23rd century. Beyond AGI lies the potential for superintelligent AI, an intelligence surpassing human cognitive abilities across virtually every domain.

Superintelligent AI could revolutionize every aspect of human life, from science and medicine to economics and governance. It could solve currently intractable problems, such as curing diseases, addressing climate change, and even answering fundamental questions about the nature of the universe. However, the development of superintelligent AI also raises significant risks, including the potential loss of human control over such systems and the possibility of unforeseen consequences.

By 3024, conversations might revolve around the role of superintelligent AI in society, including how it is integrated into human life, its impact on employment and social structures, and the ethical considerations of creating and controlling such powerful entities. The relationship between humans and AI could involve collaboration, with AI augmenting human capabilities, or it could evolve into more complex dynamics, such as the potential for AI to develop its own goals and priorities.

The creation of superintelligent AI also raises profound ethical and philosophical questions, particularly regarding the nature of consciousness, intelligence, and morality. If AI systems become self-aware or capable of experiencing emotions, how should they be treated? What rights and responsibilities would such entities have, and how would they fit into human society?

Conversations in 3024 might explore these questions and the potential for AI to challenge our understanding of what it means to be human. If AI systems can think, feel, and create in ways that surpass human capabilities, how will this affect our sense of identity and purpose? Will humans become obsolete, or will AI be seen as an extension of human creativity and innovation?

The ethical considerations of superintelligent AI also extend to control and governance issues. Ensuring that AI systems align with human values and priorities will be a critical challenge, particularly as these systems become more autonomous and powerful. The potential for AI to be used for malicious purposes, whether by individuals, corporations, or governments, will also be a topic of concern, prompting discussions about the regulation and oversight of AI technologies.

The Integration of Humans and Machines

As technology advances, the boundaries between humans and machines could become increasingly indistinct. Brain-computer interfaces (BCIs), which allow direct communication between the brain and external devices, are already being developed for medical applications, such as restoring movement to individuals with paralysis or enabling communication for those with severe disabilities. In the future, BCIs could become a standard tool for enhancing cognitive abilities, accessing information, and controlling technology with the power of thought.

In 3024, conversations might focus on integrating BCIs and other advanced technologies into daily life. How will these technologies change how we interact with the world, each other, and ourselves? Will they enhance our abilities, creativity, and productivity or lead to new dependence and inequality?

Integrating humans and machines also raises questions about the nature of identity and individuality. As we augment our brains with artificial intelligence, download our memories into digital storage, or even merge our consciousness with others, what will it mean to be a distinct individual? Will the self concept become fluid, or will new forms of identity and community emerge?

Developing advanced prosthetics, implants, and genetic modifications could create "cyborgs"—humans with enhanced physical and cognitive abilities. These enhancements could range from improved strength, speed, and endurance to enhanced memory, intelligence, and sensory perception. In some cases, individuals might choose to replace or augment parts of their bodies with artificial components, leading to a blending of biological and mechanical systems.

Conversations in 3024 might explore the implications of human enhancement, including the ethical and social challenges of creating a society where some individuals have access to superior abilities. Will enhanced humans have an advantage over unmodified ones, leading to new forms of inequality? How will society ensure that enhancements are available to all rather than restricted to the wealthy or privileged?

The concept of cyborgs also raises questions about the nature of humanity. What does it mean to be human if we can modify our bodies and minds so that we are no longer entirely human? Will humanity evolve into a new species, or will we retain our connection to our biological roots?

The Potential for Technological Singularity

The technological singularity refers to a hypothetical point in the future when technological growth becomes uncontrollable and irreversible, creating superintelligent AI that surpasses human intelligence. The singularity is often associated with the idea that this event could lead to rapid, exponential technological advances, fundamentally transforming society and the human experience.

In 3024, conversations might revolve around the aftermath of the singularity, assuming it has occurred. How has society adapted to the rapid changes brought about by superintelligent AI and other advanced technologies? What new forms of governance, ethics, and social structures have emerged to manage the challenges and opportunities of the singularity?

The singularity also raises questions about the future of human evolution. Will humans continue to evolve alongside technology, or will our creations surpass us? What role will humans play in a world dominated by superintelligent AI, and how will we preserve our values and priorities?

The singularity presents both risks and opportunities for the future of humanity. On one hand, the rapid advancement of technology could lead to unprecedented progress in science, medicine, and quality of life. On the other hand, the potential loss of control over superintelligent AI and other technologies could pose existential threats to humanity.

Conversations in 3024 might focus on the

strategies used to navigate the singularity, including developing safeguards, ethical frameworks, and international cooperation to manage the risks. The opportunities the singularity presents, such as the potential for solving global challenges and expanding human capabilities, will also be a topic of interest as society grapples with the balance between innovation and safety.

The Future of Space Exploration and Colonization

The exploration and colonization of space represent some of the most ambitious goals for humanity's future. As Earth becomes increasingly interconnected and global challenges such as climate change and resource depletion intensify, space offers new frontiers for exploration, habitation, and the expansion of human civilization.

The Colonization of the Solar System

By 3024, humanity may have established permanent colonies on planets, moons, and asteroids within the solar system. With its relatively similar day length and potential for terraforming, Mars is a prime candidate for colonization, as are the moons of Jupiter and Saturn, such as Europa and Titan, which may harbor subsurface oceans and the potential for life.

Conversations in 3024 might focus on the challenges and successes of space colonization, including the technological, logistical, and ethical considerations of establishing human settlements on other worlds. How have humans adapted to life in space, both physically and psychologically? What new forms of governance, culture, and economy have emerged in these extraterrestrial communities?

Colonizing the solar system also raises questions about the long-term sustainability of human civilization. How have space colonies managed resources, such as water, air, and food, in environments where these are scarce? What role do these colonies play in the broader context of human civilization, and how have they contributed to advancing science, technology, and exploration?

Terraforming, modifying a planet's environment to make it more Earth-like and suitable for human habitation, could be a key goal for space colonization. With its thin atmosphere and cold temperatures, Mars is often cited as a candidate for terraforming, which would involve thickening the atmosphere, warming the planet, and introducing water and vegetation.

In 3024, conversations might explore the progress and challenges of terraforming efforts, as well as the ethical implications of altering the environment of another planet. Should humanity have the right to terraform other worlds, potentially disrupting existing ecosystems or life forms? What responsibilities do we have to preserve the natural state of other planets, and how do we balance the needs of human colonists with the protection of extraterrestrial environments?

Creating habitable environments might also extend beyond planets to include artificial habitats, such as space stations, orbital cities, or O'Neill cylinders —massive rotating structures that simulate gravity and provide a self-sustaining environment for large populations. The development of such habitats could expand human civilization beyond the limitations of planetary surfaces, creating new opportunities for exploration and habitation.

The Exploration of the Galaxy

By 3024, humanity may have developed the technology for interstellar travel, enabling the exploration of distant star systems and the search for extraterrestrial life. Advances in propulsion technology, such as fusion drives, antimatter engines, or even warp drives, could make it possible to reach other star systems within a human lifetime.

Conversations in 3024 might focus on the achievements and challenges of interstellar exploration, including the discovery of exoplanets, the potential for habitable worlds, and the search for intelligent life. How has humanity expanded its reach into the galaxy, and what new frontiers of science and exploration have been opened up by interstellar travel?

The search for life beyond Earth could also be a central topic of discussion. Have we discovered signs of life on other planets, moons, or in the atmospheres of distant exoplanets? If so, what are the implications for our understanding of biology, evolution, and the uniqueness of life on Earth? The discovery of extraterrestrial life, whether microbial or intelligent, would have profound implications for our understanding of the universe and our place within it.

Humanity's expansion beyond the solar system raises the possibility of developing a galactic civilization, with human presence spread across multiple star systems. Such a civilization could be connected by advanced communication networks, transportation systems, and shared cultural, scientific, and economic ties.

In 3024, conversations might explore the structure and organization of a galactic civilization, including the challenges of maintaining cohesion and communication across vast distances. How have humans adapted to the diversity of environments and cultures that might emerge in different star systems? What new governance, ethics, and philosophy forms have arisen in response to humanity's expansion across the galaxy?

The concept of a galactic civilization also raises questions about the potential for encountering other intelligent species. If we discover advanced extraterrestrial civilizations, how will we interact with them, and how will this impact our understanding of intelligence, culture, and morality? The potential for cooperation, conflict, or coexistence with other species will be a central topic of discussion in a future where humanity has expanded into the galaxy.

The Evolution of Human Biology: Genetic Engineering and Biotechnology

Genetic engineering, biotechnology, and medical advances could lead to profound changes in human biology over the next thousand years. The ability to modify the human genome, enhance physical and cognitive skills, and prevent or reverse aging could redefine what it means to be human.

Genetic Engineering and Human Enhancement

By 3024, the ability to modify the human genome may be routine, allowing individuals to customize their genetic makeup before birth or even throughout their lives. Genetic engineering could eliminate inherited diseases, enhance physical and cognitive abilities, and modify traits such as appearance, intelligence, and longevity.

Conversations in 3024 might focus on the implications of widespread genetic engineering, including the ethical considerations of designing and enhancing humans. What impact has genetic engineering had on society, diversity, and equality? How have individuals and cultures responded to the ability to customize the human genome, and what new social norms and values have emerged?

Customizing the human genome also raises questions about the long-term consequences of genetic engineering. How have these modifications affected human evolution, and what new traits or abilities have emerged? The potential for unintended consequences, such as the emergence of new diseases or the loss of genetic diversity, will also be a topic of concern.

The ability to modify the human genome could also lead to the development of new human species or subspecies, each adapted to different environments, lifestyles, or goals. These new species might be designed for life in space, with enhanced radiation resistance, lower oxygen requirements, or the ability to thrive in microgravity. Others might be designed for specific cognitive or physical tasks, such as advanced problem-solving, artistic creativity, or athletic performance.

Conversations in 3024 might explore the implications of the emergence of a new human species, including the potential for social, cultural, and political divisions. How have these new species integrated into society, and what relationships have developed between different human groups? The concept of humanity itself may evolve as new species challenge our understanding of identity, rights, and human nature.

Creating a new human species also raises ethical questions about the rights and responsibilities of those involved in genetic engineering. Who has the authority to design or modify human beings, and what safeguards are in place to prevent abuse or exploitation? The potential for genetic inequality, discrimination, or even conflict between different species will be a central concern in a future where genetic engineering is commonplace.

Biotechnology and the Extension of Life

Advances in biotechnology, regenerative medicine, and genetic engineering could extend human lifespan, potentially achieving biological immortality. The ability to prevent or reverse aging, repair damaged tissues, and eliminate age-related diseases could enable individuals to live for centuries or even millennia.

Conversations in 3024 might focus on the achievements and challenges of life extension, including the impact on society, culture, and the individual. How has the extension of life affected the way people view time, relationships, and personal growth? What new challenges have emerged from living for centuries, such as the potential for overpopulation, resource depletion, or social stagnation?

The quest for immortality also raises philosophical and ethical questions about the nature of life and death. If death is no longer an inevitable part of life, how will this affect our understanding of purpose, meaning, and the human experience? The potential for immortality could lead to new forms of spirituality, ethics, and existential reflection as individuals and societies grapple with the implications of eternal life.

Regenerative medicine, which involves using stem cells, tissue engineering, and gene therapy to repair or replace damaged tissues and organs, could significantly extend life. The ability to regenerate tissues, reverse aging, and restore function to aged organs could enable individuals to maintain their health and vitality well into old age.

In 3024, conversations might explore advances in regenerative medicine, including developing new therapies, the ethical considerations of rejuvenation, and the impact on healthcare and society. How have these therapies transformed the aging process, and what new possibilities have emerged for maintaining health and well-being throughout life?

The potential for rejuvenation also raises questions about the balance between extending life and preserving the quality of life. How have individuals and societies navigated the challenges of living longer, including the potential for boredom, loss of purpose, or social isolation? The pursuit of rejuvenation may lead to new forms of community, social structures, and cultural practices that support individuals' well-being over extended lifespans.

The Future of Reproduction and Human Development

Advances in reproductive technology could lead to the development of artificial wombs, enabling gestation outside the human body. This technology could eliminate the risks and limitations of natural pregnancy, allowing for greater control over the development of embryos and the ability to customize traits through genetic engineering.

Conversations in 3024 might explore the implications of artificial wombs and designer babies, including the ethical considerations of human reproduction and the potential impact on family structures, gender roles, and society. How have these technologies changed how people view parenthood, childbirth, and the development of children? What new social norms and legal frameworks have emerged to govern reproductive technology?

The ability to design and customize babies through genetic engineering could also raise questions about the future of human diversity and the potential for eugenics. How have societies ensured that these technologies are used responsibly and ethically, and what safeguards are in place to prevent discrimination or coercion? The potential for creating a genetically enhanced elite or exacerbating existing social inequalities will be a central concern in a future where reproductive technology is highly advanced.

Advances in biotechnology and genetic engineering could also lead to new forms of human development, including the ability to accelerate or modify growth, cognitive development, and aging. The ability to control and enhance development at every stage of life could lead to new possibilities for education, creativity, and personal growth.

In 3024, conversations might focus on the impact of these technologies on the human experience, including the potential for new forms of learning, self-expression, and personal development. How have these technologies changed how people view childhood, adolescence, and adulthood? What new opportunities and challenges have emerged due to the ability to shape human development?

The evolution of human development also raises questions about the future of education and socialization. How have educational systems adapted to the new possibilities of enhanced learning and cognitive development? What new forms of community and social interaction have emerged due to changes in the human lifespan and developmental stages?

The Future of Energy and the Environment

The future of energy and the environment will be critical topics of conversation in the year 3024 as humanity grapples with the challenges of sustaining life on Earth and beyond. Advances in energy technology, environmental conservation, and climate management could shape the future of human civilization and the planet.

Sustainable Energy and Climate Management

By 3024, humanity may have fully transitioned to renewable energy sources, such as solar, wind, and fusion power, eliminating reliance on fossil fuels and reducing the environmental impact of energy production. Advances in energy storage, transmission, and efficiency could enable the widespread use of clean energy, supporting the needs of a growing global population and the expansion of human civilization into space.

Conversations in 3024 might focus on the achievements and challenges of the energy transition, including the impact on economies, industries, and the environment. How have renewable energy technologies transformed how we produce and consume energy, and what new possibilities have emerged for sustainable development? The potential for energy abundance and the need for responsible resource management will be central topics of discussion.

The transition to renewable energy also raises questions about the long-term sustainability of human civilization. How have societies managed the balance between energy production and environmental conservation? What new forms of governance and cooperation have emerged to address global challenges, such as climate change, resource depletion, and biodiversity loss?

Climate engineering, the deliberate manipulation of the Earth's climate system, could play a central role in managing the impacts of climate change and restoring the environment. Technologies such as carbon capture and storage, solar radiation management, and large-scale reforestation could be used to stabilize the climate, reduce greenhouse gas concentrations, and protect ecosystems.

In 3024, conversations might explore the successes and challenges of climate engineering and the ethical considerations of intervening in the Earth's climate system. How have these technologies been used to address the challenges of climate change, and what impact have they had on the environment and society? Key discussion topics will be the potential for unintended consequences and the need for international cooperation and governance.

Restoring degraded ecosystems and protecting biodiversity could also be central goals in 3024. Advances in environmental science, conservation biology, and biotechnology could enable the restoration of damaged habitats, the revival of extinct species, and the protection of endangered ecosystems. The future of environmental conservation will depend on balancing human needs with preserving the natural world, ensuring that future generations inherit a healthy and resilient planet.

The Exploration and Utilization of Space Resources

The exploration and utilization of space resources, such as minerals, water, and energy, could play a crucial role in supporting the expansion of human civilization into space. Mining asteroids, moons, and other celestial bodies could provide valuable resources for space colonies, reducing the need for resource extraction on Earth and enabling the construction of space infrastructure.

Conversations in 3024 might focus on the development of space mining technologies, the ethical considerations of resource extraction in space, and the potential impact on Earth and other celestial bodies. How have these technologies been used to support the expansion of human civilization, and what new opportunities and challenges have emerged? The potential for conflict over space resources and the need for international cooperation and regulation will be central discussion topics.

Exploring space resources also raises questions about the long-term sustainability of human civilization. How have societies managed the balance between resource extraction and environmental conservation on Earth and in space? What new forms of governance and cooperation have emerged to address the challenges of space exploration and resource management?

Space-based energy systems, such as solar power satellites, could provide a virtually unlimited clean energy supply for Earth and space colonies. These systems would capture solar energy in space and transmit it to Earth or other locations using microwave or laser beams, providing a reliable and sustainable energy source.

In 3024, conversations might explore the development and deployment of space-based energy systems, including such projects' technological, logistical, and ethical considerations. How have these systems been used to support the needs of a growing global population, and how have they impacted the environment and economy? The potential for energy abundance and the need for responsible resource management will be key topics of discussion.

The development of space-based energy systems also raises questions about the future of energy distribution and governance. How have societies managed the balance between energy production and consumption, and what new forms of cooperation and regulation have emerged to address energy distribution challenges on a global and interplanetary scale?

The Long-Term Sustainability of Human Civilization

As human civilization expands and develops, the balance between growth and sustainability will be a central concern. The need to provide for a growing global population while protecting the environment and ensuring the planet's long-term health will require new approaches to resource management, energy production, and environmental conservation.

Conversations in 3024 might focus on strategies to achieve this balance, including developing sustainable technologies, protecting biodiversity, and managing natural resources. How have societies navigated growth and sustainability challenges, and what new forms of governance, cooperation, and ethics have emerged to support these goals?

The balance between growth and sustainability also raises questions about the future of human civilization. How have societies adapted to the challenges of environmental change, resource depletion, and population growth? What new opportunities and challenges have emerged due to these changes, and how have they shaped the future of humanity?

The long-term sustainability of human civilization will depend not only on technological and scientific advances but also on developing new ethical frameworks and philosophies. The need to balance the needs of the present with the rights of future generations and the responsibility to protect the environment and other species will require new forms of ethical reasoning and decision-making.

In 3024, conversations might explore the role of ethics and philosophy in sustainability, including the development of new ethical principles, the impact of these principles on policy and governance, and the potential for global cooperation to address shared challenges. The future of sustainability will depend on integrating ethical considerations into decision-making, ensuring that the choices made today support the well-being of future generations.

The role of ethics and philosophy in sustainability also raises questions about the future of human values and priorities. How have societies navigated the challenges of environmental change, resource depletion, and population growth, and what new forms of ethics and philosophy have emerged to support these goals? The potential for new forms of spirituality, morality, and meaning in a world shaped by sustainability will be central topics of discussion.

The Social and Ethical Implications of Future Developments

The advancements in technology, space exploration, human biology, and environmental sustainability discussed in this chapter will have profound social and ethical implications. These developments will challenge our understanding of what it means to be human, how we interact with each other and the world, and how we navigate the complex and interconnected challenges of the future.

The Evolution of Social Structures and Governance

The rapid advancement of technology, particularly in AI, biotechnology, and space exploration, will transform society. Integrating AI into everyday life, enhancing human biology, and colonizing space will lead to new forms of social organization, governance, and culture.

Conversations in 3024 might focus on the evolution of social structures in response to these changes, including developing new forms of government, economy, and community. How have societies adapted to the challenges and opportunities of technological advancement, and what new forms of cooperation and conflict have emerged?

The impact of technology on society also raises questions about the future of human values and priorities. How have individuals and communities navigated the challenges of rapid technological change, and what new forms of ethics, philosophy, and spirituality have emerged to support these goals? The potential for new forms of identity, culture, and social interaction in a world shaped by technology will be central discussion topics.

The expansion of human civilization into space and the need to address global challenges such as climate change and resource depletion will require new forms of governance and international cooperation. The development of space law, the regulation of space resources, and the management of shared global resources will be critical to the long-term sustainability of human civilization.

In 3024, conversations might explore the challenges and successes of global governance, including the development of new institutions, the role of international cooperation, and the potential for conflict or collaboration between different nations and cultures. How have societies navigated the challenges of space exploration and global resource management, and what new forms of governance have emerged to support these goals?

The governance of space and global challenges also raise questions about the future of democracy, human rights, and social justice. How have societies ensured that the benefits of space exploration and technological advancement are shared equitably, and what new forms of governance and ethics have emerged to support these goals? The potential for new forms of global cooperation, conflict, and management in a world shaped by space exploration and technological advancement will be central topics of discussion.

The Ethical Considerations of Human Enhancement and Longevity

The ability to modify the human genome and enhance physical and cognitive skills raises profound ethical questions about the future of humanity. The potential for creating genetically enhanced individuals, new human species, or even artificial intelligence systems challenges our understanding of identity, rights, and the nature of being human.

Conversations in 3024 might focus on the ethical considerations of human enhancement, including the potential for inequality, discrimination, and exploitation. How have societies navigated the challenges of genetic engineering and human enhancement, and what new forms of ethics and governance have emerged to support these goals?

The ethics of human enhancement also raise questions about the future of human diversity and the potential for new forms of social organization and identity. How have individuals and communities navigate the challenges of genetic engineering and human enhancement, and what new forms of ethics, philosophy, and spirituality have emerged to support these goals? The potential for new forms of identity, culture, and social interaction in a world shaped by human enhancement will be central topics of discussion. The potential for extending human lifespan or achieving biological immortality raises profound ethical questions about the nature of life and death, the balance between quality and quantity of life, and the impact on society and the environment.

Conversations in 3024 might explore the ethical considerations of longevity and immortality, including the potential for overpopulation, resource depletion, and social inequality. How have societies navigated the challenges of life extension, and what new forms of ethics, philosophy, and spirituality have emerged to support these goals?

The ethics of longevity and immortality also raise questions about the future of human values and priorities. How have individuals and communities navigated the challenges of extended life, and what new forms of ethics, philosophy, and spirituality have emerged to support these goals? The potential for new forms of identity, culture, and social interaction in a world shaped by longevity and immortality will be central topics of discussion.

The Future of Human Values and Ethics

The rapid advancement of technology, combined with expanding human civilization into space and the potential for human enhancement and longevity, will require new forms of ethical reasoning and decision-making. The need to balance individuals' rights with society's needs, the protection of the environment, and the long-term sustainability of human civilization will require new forms of ethics and philosophy.

In 3024, conversations might explore the

evolution of ethics in response to these changes, including the development of new ethical principles, the impact of these principles on policy and governance, and the potential for global cooperation to address shared challenges. How have societies navigated the challenges of technological change, and what new forms of ethics and philosophy have emerged to support these goals?

The evolution of ethics in response to technological change also raises questions about the future of human values and priorities. How have individuals and communities navigated the challenges of rapid technological change, and what new forms of ethics, philosophy, and spirituality have emerged to support these goals? The potential for new forms of identity, culture, and social interaction in a world shaped by technology will be central discussion topics.

The challenges and opportunities of the future will require new forms of philosophy and spirituality to guide individuals and societies through the complexities of life in the year 3024. The need to balance technological advancement with ethical considerations, the protection of the environment, and the long-term sustainability of human civilization will require new forms of moral reasoning and decision-making.

Conversations in 3024 might explore the role of philosophy and spirituality in the future, including the development of new ethical principles, the impact of these principles on policy and governance, and the potential for global cooperation to address shared challenges. How have societies navigated the challenges of technological change, and what new forms of ethics and philosophy have emerged to support these goals?

The role of philosophy and spirituality in the future also raises questions about the evolution of human values and priorities. How have individuals and communities navigated the challenges of rapid technological change, and what new forms of ethics, philosophy, and spirituality have emerged to support these goals? The potential for new forms of identity, culture, and social interaction in a world shaped by technology will be central discussion topics.

Conclusion

Summary of Key Points

The year 3024 will likely be characterized by profound technological advancements, space exploration, human biology, and environmental sustainability. These developments will reshape human civilization and challenge our understanding of what it means to be human.

The rise of superintelligent AI, the integration of humans and machines, the colonization of space, the extension of human lifespan, and the exploration of new forms of energy and environmental conservation will be central topics of conversation. These advancements will raise significant social, ethical, and philosophical questions.

The evolution of social structures, governance, and ethics in response to these changes will be critical to the long-term sustainability of human civilization. New forms of cooperation, conflict, and identity will emerge as humanity navigates the challenges and opportunities of the future.

The role of philosophy, ethics, and spirituality

in guiding individuals and societies through the complexities of life in 3024 will be essential. The need to balance technological advancement with ethical considerations, the protection of the environment, and the long-term sustainability of human civilization will require new forms of moral reasoning and decision-making.

The Importance of Continued Inquiry and Exploration

The study of the future is an ongoing area of inquiry that challenges our understanding of human civilization, technology, and the environment. Continued exploration and inquiry are essential for advancing our knowledge of these topics and for developing strategies to navigate the challenges and opportunities of the future.

Advances in technology, space exploration, human biology, and environmental sustainability offer new possibilities for the future of human civilization. However, these advancements also raise significant ethical and social considerations that must be addressed as we explore the potential for life in 3024.

The Future as a Source of Wonder and Inspiration

The exploration of the future is a source of wonder and inspiration, offering new perspectives on the nature of human civilization, technology, and the environment. It challenges us to consider human progress's possibilities and limitations and explore the potential for extending the boundaries of life and civilization.

The study of the future invites us to reflect on the

value of life, the meaning of technological advancement, and the impact of human civilization on the planet and the universe. It encourages us to consider our choices' ethical and philosophical implications and explore new frameworks for understanding the relationship between humanity, technology, and the environment.

Whether or not we can predict the specifics of life in 3024, exploring the future highlights the importance of promoting ethical reasoning, sustainability, and cooperation throughout the lifespan of human civilization. It reminds us that the quest for knowledge and progress is not just about advancing technology but ensuring that our choices support future generations' well-being and the long-term sustainability of life on Earth and beyond.

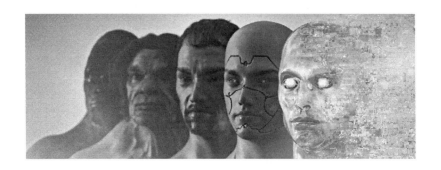

CHAPTER 10: EPILOGUE: RECONCILING FAITH AND SCIENCE

This final chapter explores the harmony between Christian faith and philosophical inquiries. It delves into the concept that religion, while it may function as a placebo, possesses an immense power that can transcend rational arguments. Through this lens, we aim to find peace by embracing the duality of faith and reason, understanding the role of religion in human life, and integrating these two seemingly disparate parts of ourselves.

Religion can be viewed as a practical, powerful, and

transformative placebo. This perspective allows us to maintain philosophical rigor while acknowledging the profound impact of faith on our lives and the lives of others. It respects both the spiritual and intellectual aspects of our being, creating a pathway to inner peace.

The placebo effect is a well-documented phenomenon in medicine where patients experience real improvements in health or well-being after receiving treatments with no therapeutic value. The power of the placebo effect lies in the mind's ability to influence the body's physiological responses, leading to measurable changes in health outcomes. This effect extends beyond physical ailments, influencing psychological and emotional well-being as well.

Belief is at the core of the placebo effect. When individuals believe a treatment will work, their expectations can trigger psychological and physiological responses that contribute to healing. The brain releases endorphins and other chemicals to alleviate pain, reduce stress, and promote well-being. This illustrates the profound connection between mind and body, where belief can influence physical reality.

In religion, belief plays a similar role. Faith in a higher power, religious teachings, or the efficacy of prayer can bring tangible benefits, including a sense of purpose, inner peace, moral guidance, and a supportive community. Just as a placebo can heal the body, religious belief can heal the soul, providing comfort, hope, and resilience in life's challenges.

Rituals and symbols are central to both the placebo effect

and religious practices. In medicine, receiving a placebo—whether in the form of a pill, a procedure, or a therapeutic ritual—can enhance belief in its efficacy, amplifying the placebo effect. Similarly, religious traditions such as prayer, meditation, communion, or pilgrimage reinforce belief and create a sense of connection to the divine. These rituals and symbols serve as focal points for faith, grounding abstract beliefs in tangible actions and experiences, and providing structure and meaning to help navigate life's complexities with a sense of order and purpose.

The placebo effect is often enhanced by a supportive environment, such as trust in a healthcare provider or the encouragement of a community. In religion, the communal aspect is equally significant. Belonging to a religious community provides social support, shared values, and collective practices that reinforce belief and promote psychological well-being. Religious communities offer a sense of belonging and identity, creating safe spaces for individuals to express their beliefs, share their struggles, and find solace among those who share their faith. This communal support can be particularly powerful in times of crisis, providing strength and encouragement.

Viewing religion as a powerful placebo offers a framework for understanding faith that aligns with philosophical skepticism. From this perspective, religious beliefs and practices are not necessarily about objective truth but about the subjective experience of finding meaning in life. Religion, like a placebo, can have real and lasting effects on mental and emotional well-being, even if it does not correspond to empirical reality.

This approach allows us to appreciate the value of Christian faith without needing to justify it solely through rational arguments. Rather than focusing on whether religious claims are objectively true, we can focus on how they enrich our lives, provide moral guidance, and foster a connection to something greater than ourselves.

Acknowledging religion as a placebo does not diminish its importance or validity; instead, it underscores the compatibility of faith and reason. As philosophers, we can recognize the limits of human knowledge and the role of belief in shaping our understanding of the world. In this context, faith becomes a complementary force to reason, offering insights and experiences that reason alone cannot provide.

This perspective allows for intellectual integrity while embracing the emotional and spiritual benefits of Christian faith. It encourages philosophical inquiry with an open mind, questioning and exploring different perspectives, while also acknowledging the unique role of faith in our lives. Like a powerful placebo, religion has the potential to transform lives, inspiring acts of compassion, forgiveness, and selflessness. It offers comfort in sorrow, hope in despair, and strength in weakness.

Reconciling the duality of faith and reason requires an understanding of their respective strengths and limitations. It also involves accepting the inherent tensions between them, recognizing that they are not necessarily opposing forces but complementary aspects of the human experience. Philosophers often embrace doubt and uncertainty, questioning assumptions, challenging established beliefs, and exploring multiple perspectives. This intellectual curiosity can sometimes lead to unease, especially concerning deeply held religious beliefs. However, doubt and uncertainty need not be viewed as threats to faith. Instead, they can be opportunities for growth and deeper understanding, leading to a more nuanced and resilient faith that is open to inquiry and adaptable to life's challenges.

While philosophy encourages questioning, faith provides a foundation for navigating life's uncertainties. It offers trust and confidence in the face of the unknown, allowing us to move forward even when answers are not readily available. This trust is not blind but acknowledges the limitations of human knowledge and the need for something greater to guide and sustain us.

Faith and reason can work together to create a balanced approach to life. Reason helps critically evaluate beliefs, while faith provides the courage to embrace them, even when they cannot be fully justified through logic alone. This balance allows for integrity and authenticity in philosophical inquiries and religious practices.

Integrating faith and reason involves adopting a holistic approach to truth, recognizing that truth can take many forms, including empirical, moral, spiritual, and experiential. This perspective allows us to appreciate the insights offered by both philosophy and religion without feeling compelled to prioritize one over the other. Philosophy might provide a framework for understanding reality, ethics, and human existence, while religion offers insights into love, compassion, and the divine. Both perspectives contribute to a fuller understanding of the human experience, necessary for navigating life's complexities.

Faith and reason are complementary forces, each contributing to a more complete and fulfilling life. Reason provides the tools for critical thinking, analysis, and problem-solving, while faith offers the emotional and spiritual support needed to cope with life's uncertainties and challenges. Embracing both leads to a richer and more balanced worldview.

Finding peace in the duality of faith and reason requires acceptance and integration. Acceptance involves acknowledging the inherent tensions between faith and reason without needing to resolve them completely. Integration involves finding ways to harmonize these aspects of identity, allowing them to coexist and complement each other. This process can lead to inner peace, where conflicting beliefs are no longer seen as divisions but as parts of a whole.

Living with paradox is part of the journey to inner peace. Life is full of contradictions and uncertainties, and the tension between faith and reason is just one example. Embracing paradox allows us to find peace amid complexity, accepting that some questions may never have clear answers. This acceptance is not a weakness but a strength, offering a more nuanced and resilient approach to life's challenges.

Religion, whether viewed as a placebo or a source of transcendent truth, plays a significant role in society and personal life. It shapes values, behaviors, and social structures, providing a framework for understanding the world and our place within it. Religion can unite people, creating a sense of belonging and shared purpose. Religious communities offer support, guidance, and a sense of identity, helping individuals navigate life's complexities. By participating in a religious community, we can find connection and fellowship that enhance spiritual and emotional well-being.

Religion also plays a crucial role in shaping moral and ethical behavior. Religious teachings offer guidelines for living a good and meaningful life, providing principles of justice, compassion, and integrity. These teachings can serve as a foundation for personal and social ethics, helping individuals make decisions that align with their values and beliefs.

On a personal level, religion provides a deep sense of spiritual and emotional fulfillment. It offers a way to connect with the divine, explore the mysteries of existence, and find meaning amid suffering and uncertainty. Embracing the spiritual and emotional benefits of faith allows us to find peace in its ability to nourish the soul and guide life's journey.

Religion is about a personal relationship with the divine, unique to each individual. This relationship can take many forms, from a deep sense of connection and trust to moments of doubt and questioning. By embracing this relationship, we can find peace in the knowledge that faith is a journey that evolves and grows over time. This personal relationship allows us to explore faith on our terms, integrating philosophical insights and questions into spiritual practice. Faith is not about having all the answers but about seeking a deeper understanding of ourselves, the world, and the divine.

In conclusion, finding peace in the harmony of faith and reason involves recognizing the transformative power of religion as a placebo and integrating faith with philosophical inquiry. By appreciating both the emotional and spiritual support of religion and the intellectual challenges of philosophy, we can live a meaningful and intellectually fulfilling life where beliefs and questions coexist in dynamic and enriching dialogue.

As we continue on this journey, remember that faith and reason are valuable tools for navigating life's complexities. Embracing both leads to a rich life filled with meaning, purpose, and understanding. The path to inner peace is not a destination but an ongoing exploration, reflection, and growth. By staying open to new insights and experiences, we deepen our understanding of ourselves, our faith, and the world around us.

REFERENCES: CHAPTER 1

1. Greene, B. (2004). *The Fabric of the Cosmos: Space, Time, and the Texture of Reality.* Alfred A. Knopf.
2. Hawking, S. (1988). *A Brief History of Time: From the Big Bang to Black Holes.* Bantam Books.
3. Ellis, G. F. R., & Silk, J. (2014). "Scientific Method: Defend the Integrity of Physics." *Nature,* 516(7531), 321-323.
4. Carroll, S. M. (2014). "In What Sense Is the Early Universe Fine-Tuned?" *Foundations of Physics,* 44, 7-29.
5. Planck Collaboration. (2016). "Planck 2015 results - XIII. Cosmological parameters." *Astronomy & Astrophysics,* 594, A13. https://doi.org/10.1051/0004-6361/201525830
6. Riess, A. G., et al. (1998). "Observational Evidence from Supernovae for an Accelerating Universe and a Cosmological Constant." *The Astronomical Journal,* 116(3), 1009-1038. https://doi.org/10.1086/300499
7. Schneider, R., et al. (2002). "First stars,

supernovae and the intergalactic medium." *Astronomy & Astrophysics*, 384(1), 58-68. https://doi.org/10.1051/0004-6361:20011788

8. Carroll, S. M. (2001). "The Cosmological Constant." *Living Reviews in Relativity*, 4(1), 1. https://doi.org/10.12942/lrr-2001-1

9. Tegmark, M., et al. (2004). "Cosmological parameters from SDSS and WMAP." *Physical Review D*, 69(10), 103501. https://doi.org/10.1103/PhysRevD.69.103501

10. Planck Collaboration. (2018). "Planck 2018 results. VI. Cosmological parameters." *Astronomy & Astrophysics*, 641, A6. https://doi.org/10.1051/0004-6361/201833910

11. Spergel, D. N., et al. (2003). "First-Year Wilkinson Microwave Anisotropy Probe (WMAP) Observations: Determination of Cosmological Parameters." *The Astrophysical Journal Supplement Series*, 148(1), 175. https://doi.org/10.1086/377226

REFERENCES: CHAPTER 2

1. Descartes, R. (1641). *Meditations on First Philosophy.* Translated by J. Cottingham. Cambridge University Press.
2. Dennett, D. C. (1991). *Consciousness Explained.* Little, Brown and Co.
3. Metzinger, T. (2003). "Phenomenal Consciousness: The First-Person Perspective." *Scholarly Articles on Consciousness,* Oxford University Press.
4. Chalmers, D. J. (1995). "Facing Up to the Problem of Consciousness." *Journal of Consciousness Studies,* 2(3), 200-219.
5. Planck Collaboration. (2016). "Planck 2015 results - XIII. Cosmological parameters." *Astronomy & Astrophysics,* 594, A13. https://doi.org/10.1051/0004-6361/201525830
6. Riess, A. G., et al. (1998). "Observational Evidence from Supernovae for an Accelerating Universe and a Cosmological Constant." *The Astronomical Journal,* 116(3), 1009-1038. https://doi.org/10.1086/300499
7. Schneider, R., et al. (2002). "First stars,

supernovae and the intergalactic medium." *Astronomy & Astrophysics*, 384(1), 58-68. https://doi.org/10.1051/0004-6361:20011788

8. Carroll, S. M. (2001). "The Cosmological Constant." *Living Reviews in Relativity*, 4(1), 1. https://doi.org/10.12942/lrr-2001-1

9. Tegmark, M., et al. (2004). "Cosmological parameters from SDSS and WMAP." *Physical Review D*, 69(10), 103501. https://doi.org/10.1103/PhysRevD.69.103501

10. Planck Collaboration. (2018). "Planck 2018 results. VI. Cosmological parameters." *Astronomy & Astrophysics*, 641, A6. https://doi.org/10.1051/0004-6361/201833910

11. Spergel, D. N., et al. (2003). "First-Year Wilkinson Microwave Anisotropy Probe (WMAP) Observations: Determination of Cosmological Parameters." *The Astrophysical Journal Supplement Series*, 148(1), 175. https://doi.org/10.1086/377226

REFERENCES: CHAPTER 3

1. Vilenkin, A. (2006). *Many Worlds in One: The Search for Other Universes.* Hill and Wang.
2. Penrose, R. (2010). *Cycles of Time: An Extraordinary New View of the Universe.* Knopf.
3. Borde, A., Guth, A. H., & Vilenkin, A. (2003). "Inflationary Spacetimes Are Not Past-Complete." *Physical Review Letters,* 90(15), 151301.
4. Linde, A. (2007). "Inflationary Cosmology." *Reports on Progress in Physics,* 70(6), 925-986.
5. Alpher, R. A., & Herman, R. (1948). "Evolution of the Universe." *Nature,* 162(4132), 774-775. https://doi.org/10.1038/162774b0
6. Penrose, R. (1989). *The Emperor's New Mind: Concerning Computers, Minds, and the Laws of Physics.* Oxford University Press. https://doi.org/10.1063/1.2828692
7. Zeldovich, Y. B., & Novikov, I. D. (1983). *Relativistic Astrophysics, Vol. 2: The Structure and Evolution of the Universe.* University of Chicago Press. https://doi.org/10.1086/161349

8. Guth, A. H. (1981). "Inflationary Universe: A Possible Solution to the Horizon and Flatness Problems." *Physical Review D,* 23(2), 347-356. https://doi.org/10.1103/PhysRevD.23.347

9. Hawking, S., & Ellis, G. F. R. (1973). *The Large Scale Structure of Space-Time.* Cambridge University Press. https://doi.org/10.1017/CBO9780511524646

10. Lemaître, G. (1931). "The Beginning of the World from the Point of View of Quantum Theory." *Nature,* 127(3210), 706. https://doi.org/10.1038/127706b0

REFERENCES: CHAPTER 4

1. Einstein, A. (1916). *Relativity: The Special and General Theory.* H. Holt and Company.
2. Penrose, R. (2004). *The Road to Reality: A Complete Guide to the Laws of the Universe.* Jonathan Cape.
3. Will, C. M. (2014). "The Confrontation between General Relativity and Experiment." *Living Reviews in Relativity,* 17, 4.
4. Norton, J. (2000). "What Was Einstein's Principle of Equivalence?" *Studies in History and Philosophy of Science Part B: Studies in History and Philosophy of Modern Physics,* 31(2), 135-157.

REFERENCES: CHAPTER 5

1. Feynman, R. P., Leighton, R. B., & Sands, M. (1965). *The Feynman Lectures on Physics, Vol. III: Quantum Mechanics.* Addison-Wesley.
2. Rovelli, C. (2017). *Reality Is Not What It Seems: The Journey to Quantum Gravity.* Penguin Books.
3. Zeilinger, A. (1999). "Experiment and the Foundations of Quantum Physics." *Reviews of Modern Physics,* 71(2), S288.
4. Leggett, A. J. (2002). "Testing the Limits of Quantum Mechanics: Motivation, State of Play, Prospects." *Journal of Physics: Condensed Matter,* 14(15), R415.
5. Planck, M. (1901). "On the Law of Distribution of Energy in the Normal Spectrum." *Annalen der Physik,* 4(3), 553-563. doi:10.1002/andp.19013090310.
6. Einstein, A. (1905). "On a Heuristic Viewpoint Concerning the Production and Transformation of Light." *Annalen der Physik,* 17(6), 132-148. doi:10.1002/andp.19053220607.

7. Bohr, N. (1913). "On the Constitution of Atoms and Molecules." *Philosophical Magazine*, 26(151), 1-25. doi:10.1080/14786441308634955.

8. de Broglie, L. (1924). "Recherches sur la théorie des quanta" [Research on the Quantum Theory]. *Annales de Physique*, 10(3), 22-128.

9. Heisenberg, W. (1927). "Über den anschaulichen Inhalt der quantentheoretischen Kinematik und Mechanik" [On the Perceptual Content of Quantum Theoretical Kinematics and Mechanics]. *Zeitschrift für Physik*, 43, 172-198. doi:10.1007/BF01397280.

10. Schrödinger, E. (1926). "Quantisierung als Eigenwertproblem" [Quantization as an Eigenvalue Problem]. *Annalen der Physik*, 79(8), 361-376. doi:10.1002/andp.19263861802.

11. Born, M. (1926). "Zur Quantenmechanik der Stoßvorgänge" [On the Quantum Mechanics of Collision Processes]. *Zeitschrift für Physik*, 37, 863-867. doi:10.1007/BF01397477.

12. Bohr, N. (1935). "Can Quantum-Mechanical Description of Physical Reality Be Considered Complete?" *Physical Review*, 48(8), 696-702. doi:10.1103/PhysRev.48.696.

13. Bell, J. S. (1964). "On the Einstein Podolsky Rosen Paradox." *Physics Physique Физика*, 1(3), 195-200. doi:10.1103/

PhysicsPhysiqueFizika.1.195.

14. Aspect, A., Dalibard, J., & Roger, G. (1982). "Experimental Test of Bell's Inequalities Using Time-Varying Analyzers." *Physical Review Letters*, 49(25), 1804-1807. doi:10.1103/PhysRevLett.49.1804.

15. Landauer, R. (1987). "Tunneling Phenomena in Solids." *IBM Journal of Research and Development*, 1, 223-231.

16. Feynman, R. P. (1965). *The Character of Physical Law*. MIT Press.

17. Nielsen, M. A., & Chuang, I. L. (2010). *Quantum Computation and Quantum Information* (10th ed.). Cambridge University Press.

18. Shor, P. W. (1994). "Algorithms for Quantum Computation: Discrete Logarithms and Factoring." *Proceedings of the 35th Annual Symposium on Foundations of Computer Science*, 124-134. doi:10.1109/SFCS.1994.365700.

19. Bennett, C. H., & Brassard, G. (1984). "Quantum Cryptography: Public Key Distribution and Coin Tossing." *Proceedings of IEEE International Conference on Computers, Systems, and Signal Processing*, 175-179.

REFEREBCES: CHAPTER 6

1. Lane, N. (2015). *The Vital Question: Energy, Evolution, and the Origins of Complex Life.* W.W. Norton & Company.

2. Davies, P. (1999). *The Fifth Miracle: The Search for the Origin and Meaning of Life.* Simon & Schuster.

3. Martin, W., & Russell, M. J. (2007). "On the Origin of Biochemistry at an Alkaline Hydrothermal Vent." *Philosophical Transactions of the Royal Society B: Biological Sciences,* 362(1486), 1887-1925.

4. Lane, N., & Martin, W. (2010). "The Energetics of Genome Complexity." *Nature,* 467(7318), 929-934.

5. Planck, M. (1901). "On the Law of Distribution of Energy in the Normal Spectrum." *Annalen der Physik,* 4(3), 553-563. doi:10.1002/andp.19013090310.

6. Einstein, A. (1905). "On a Heuristic Viewpoint Concerning the Production and Transformation of Light." *Annalen der Physik,* 17(6), 132-148. doi:10.1002/

andp.19053220607.

7. Bohr, N. (1913). "On the Constitution of Atoms and Molecules." *Philosophical Magazine*, 26(151), 1-25. doi:10.1080/14786441308634955.

8. Miller, S. L., & Urey, H. C. (1953). "Organic Compound Synthesizers on the Primitive Earth." *Science*, 117(3046), 528-529. doi:10.1126/science.117.3046.528.

9. Gilbert, W. (1986). "The RNA World." *Nature*, 319(6055), 618. doi:10.1038/319618a0.

10. Oparin, A. I. (1924). *The Origin of Life*. Moscow Worker's Publishing House.

11. Lazcano, A., & Miller, S. L. (1999). "On the Origin of Metabolism." *Chemical Society Reviews*, 28(5), 345-351. doi:10.1039/a827899z.

12. Szostak, J. W., Bartel, D. P., & Luisi, P. L. (2001). "Synthesizing Life." *Nature*, 409(6823), 387-390. doi:10.1038/35053176.

13. Eigen, M., & Schuster, P. (1977). "The Hypercycle: A Principle of Natural Self-Organization." *Naturwissenschaften*, 64(11), 541-565. doi:10.1007/BF00450633.

14. Deamer, D. W. (1997). "The First Living Systems: A Bioenergetic Perspective." *Microbiology and Molecular Biology Reviews*, 61(2), 239-261.

15. Crick, F. H. C., & Orgel, L. E. (1973). "Directed Panspermia." *Icarus*, 19(3), 341-346.

doi:10.1016/0019-1035(73)90110-3.
16. Koshland, D. E. (2002). "The Seven Pillars of Life." *Science,* 295(5563), 2215-2216. doi:10.1126/science.1069985.
17. Joyce, G. F. (1989). "RNA Evolution and the Origins of Life." *Nature,* 338(6212), 217-224. doi:10.1038/338217a0.
18. Astrobiology: An Integrated Science Approach (2004). *NASA Astrobiology Strategy.* NASA, Washington, D.C.
19. Trevors, J. T., & Abel, D. L. (2004). "Chance and Necessity Do Not Explain the Origin of Life." *Cell Biology International,* 28(11), 729-739. doi:10.1016/j.cellbi.2004.09.004.

REFERENCES: CHAPTER 7

1. Penrose, R. (1994). *Shadows of the Mind: A Search for the Missing Science of Consciousness.* Oxford University Press.
2. Hameroff, S., & Penrose, R. (2017). *Consciousness in the Universe: A Review of the 'Orch OR' Theory.* Springer.
3. Fisher, M. P. A. (2015). "Quantum Cognition: The Possibility of Processing with Nuclear Spins in the Brain." *Annals of Physics,* 362, 593-602.
4. Tegmark, M. (2000). "The Importance of Quantum Decoherence in Brain Processes." *Physical Review E,* 61(4), 4194-4206.
5. Bell, J. S. (1964). "On the Einstein-Podolsky-Rosen Paradox." *Physics Physique Физика,* 1(3), 195-200.
6. Aspect, A., Dalibard, J., & Roger, G. (1982). "Experimental Test of Bell's Inequalities Using Time-Varying Analyzers." *Physical Review Letters,* 49(25), 1804-1807.
7. Hameroff, S., & Penrose, R. (1996). "Orchestrated Reduction of Quantum

Coherence in Brain Microtubules: A Model for Consciousness." *Mathematics and Computers in Simulation,* 40(3-4), 453-480.

8. Lambert, N., Chen, Y. N., Cheng, Y. C., Li, C. M., Chen, G. Y., & Nori, F. (2013). "Quantum Biology." *Nature Physics,* 9(1), 10-18.

9. Engel, G. S., Calhoun, T. R., Read, E. L., Ahn, T. K., Mancal, T., Cheng, Y. C., Blankenship, R. E., & Fleming, G. R. (2007). "Evidence for Wavelike Energy Transfer Through Quantum Coherence in Photosynthetic Systems." *Nature,* 446(7137), 782-786.

10. Gauger, E. M., Rieper, E., Morton, J. J., Benjamin, S. C., & Vedral, V. (2011). "Sustained Quantum Coherence and Entanglement in the Avian Compass." *Physical Review Letters,* 106(4), 040503.

11. Huelga, S. F., & Plenio, M. B. (2013). "Vibrations, Quanta and Biology." *Contemporary Physics,* 54(4), 181-207.

12. McFadden, J., & Al-Khalili, J. (2014). *Life on the Edge: The Coming of Age of Quantum Biology.* Crown Publishing.

13. Tegmark, M. (2000). "Importance of Quantum Decoherence in Brain Processes." *Physical Review E,* 61(4), 4194-4206.

14. Craddock, T. J. A., Hameroff, S. R., Ayoub, A. T., Klobukowski, M., & Tuszynski, J. A. (2015). "Anesthetic Alterations of Collective Terahertz Oscillations in Tubulin Correlate with Clinical Potency: Implications

for Anesthetic Action and Consciousness." *Journal of the Royal Society Interface,* 12(102), 20141286.

REFERENCES: CHAPTER 8

1. Sinclair, D. A., & LaPlante, M. (2019). *Lifespan: Why We Age—and Why We Don't Have To*. Atria Books.

2. De Grey, A. D. N. J., & Rae, M. (2007). *Ending Aging: The Rejuvenation Breakthroughs That Could Reverse Human Aging in Our Lifetime*. St. Martin's Press.

3. López-Otín, C., Blasco, M. A., Partridge, L., Serrano, M., & Kroemer, G. (2013). "The Hallmarks of Aging." *Cell*, 153(6), 1194-1217.

4. Campisi, J., & D'Adda Di Fagagna, F. (2007). "Cellular Senescence: When Bad Things Happen to Good Cells." *Nature Reviews Molecular Cell Biology*, 8(9), 729-740.

5. Harman, D. (1956). "Aging: A Theory Based on Free Radical and Radiation Chemistry." *Journal of Gerontology*, 11(3), 298-300.

6. Kenyon, C. (2010). "The Genetics of Ageing." *Nature*, 464(7288), 504-512.

7. López-Otín, C., Blasco, M. A., Partridge, L., Serrano, M., & Kroemer, G. (2013). "The Hallmarks of Aging." *Cell*, 153(6), 1194-1217.

8. Kirkwood, T. B. L. (2005). "Understanding the Odd Science of Aging." *Cell,* 120(4), 437-447.

9. de Magalhães, J. P., & Passos, J. F. (2018). "Stress, Cell Senescence, and Organismal Ageing." *Mechanisms of Ageing and Development,* 170, 2-9.

10. Mitchell, S. J., & Mattison, J. A. (2019). "Caloric Restriction and CR Mimetics: Can You Have Your Cake and Eat It, Too?" *Current Opinion in Clinical Nutrition & Metabolic Care,* 22(4), 267-272.

11. Campisi, J. (2013). "Aging, Cellular Senescence, and Cancer." *Annual Review of Physiology,* 75, 685-705.

12. Jaskelioff, M., et al. (2011). "Telomerase reactivation reverses tissue degeneration in aged telomerase-deficient mice." *Nature,* 469(7328), 102-106.

13. Rando, T. A., & Chang, H. Y. (2012). "Aging, Rejuvenation, and Epigenetic Reprogramming: Resetting the Aging Clock." *Cell,* 148(1-2), 46-57.

14. Scudellari, M. (2015). "Ageing research: Blood to blood." *Nature,* 517(7536), 426-429.

REFERENCES: CHAPTER 9

1. Bostrom, N. (2003). "Ethical Issues in Advanced Artificial Intelligence." *Cognitive, Emotive and Ethical Aspects of Decision Making in Humans and in Artificial Intelligence,* 2, 12-17.
2. Vinge, V. (1993). "The Coming Technological Singularity: How to Survive in the Post-Human Era." *Vision-21: Interdisciplinary Science and Engineering in the Era of Cyberspace,* 11-22.
3. Joy, B. (2000). "Why the Future Doesn't Need Us." *Wired Magazine,* April 2000.
4. Rees, M. (2003). *Our Final Hour: A Scientist's Warning: How Terror, Error, and Environmental Disaster Threaten Humankind's Future in This Century—On Earth and Beyond.* Basic Books.
5. Kaku, M. (2011). *Physics of the Future: How Science Will Shape Human Destiny and Our Daily Lives by the Year 2100.* Doubleday.
6. Kurzweil, R. (2005). *The Singularity Is Near: When Humans Transcend Biology.* Viking.
7. Silver, L. M. (1997). *Remaking Eden: Cloning*

and Beyond in a Brave New World. Avon Books.

8. Freitas Jr., R. A. (1999). *Nanomedicine, Volume I: Basic Capabilities.* Landes Bioscience.
9. Benford, G. (1999). *Deep Time: How Humanity Communicates Across Millennia.* Avon Books.
10. Dick, S. J., & Lupisella, M. (2009). *Cosmos and Culture: Cultural Evolution in a Cosmic Context.* NASA SP-4802.

REFERENCES: CHAPTER 10

1. Lewis, C. S. (1952). *Mere Christianity.* HarperOne.
2. Kierkegaard, S. (1980). *The Concept of Anxiety.* Princeton University Press.
3. Alston, W. P. (1991). "Perceiving God: The Epistemology of Religious Experience." *Journal of Philosophy,* 88(1), 93-108.
4. Plantinga, A. (2000). *Warranted Christian Belief.* Oxford University Press.
5. Vitz, P. C. (1999). *Faith of the Fatherless: The Psychology of Atheism.* Spence Publishing Company.
6. Harris, S. (2004). *The End of Faith: Religion, Terror, and the Future of Reason.* W.W. Norton & Company.
7. Barrett, J. L. (2004). *Why Would Anyone Believe in God?* AltaMira Press.
8. McGrath, A. (2004). *The Twilight of Atheism: The Rise and Fall of Disbelief in the Modern World.* Doubleday.
9. Hume, D. (1779). *Dialogues Concerning Natural*

Religion. Hafner Press.
10. James, W. (1902). *The Varieties of Religious Experience: A Study in Human Nature.* Longmans, Green, and Co.

Made in the USA
Columbia, SC
06 November 2024

105d952e-333d-493e-83a1-0cf77f060a0bR01